SITE CARPENTRY

LEVEL 2 DIPLOMA

PATRICK JOSEPH CLANCY
STEVEN LEAVERLAND

Nelson Thornes

Published in 2013 by:
Nelson Thornes Ltd
Delta Place
27 Bath Road
CHELTENHAM
GL53 7TH
United Kingdom

13 14 15 16 17 / 10 9 8 7 6 5 4 3 2 1

A catalogue record for this book is available from the British Library

ISBN 978 1 4085 2126 7

Cover photograph: tuja66/Fotolia

Page make-up by GreenGate Publishing Services, Tonbridge, Kent

Printed in Croatia by Zrinski

Note to learners and tutors

This book clearly states that a risk assessment should be undertaken and the correct PPE worn for the particular activities before any practical activity is carried out. Risk assessments were carried out before photographs for this book were taken and the models are wearing the PPE deemed appropriate for the activity and situation. This was correct at the time of going to print. Colleges may prefer that their learners wear additional items of PPE not featured in the photographs in this book and should instruct learners to do so in the standard risk assessments they hold for activities undertaken by their learners. Learners should follow the standard risk assessments provided by their college for each activity they undertake which will determine the PPE they wear.

CONTENTS

Introduction   iv

Contributors to this book   v

1 Health, Safety and Welfare in
Construction and Associated Industries   1

2 Understanding Information, Quantities
and Communication with Others   39

3 Understanding Construction Technology   67

4 Prepare and Use Carpentry and Joinery
Portable Power Tools   95

5 Carry Out First Fixing Operations   121

6 Carry Out Second Fixing Operations   181

7 Carry Out Structural Carcassing
Operations   241

8 Carry Out Maintenance to
Non-structural Carpentry Work   279

Index   311

Acknowledgements   314

INTRODUCTION

About this book

This book has been written for the Cskills Awards Level 2 Diploma in Carpentry & Joinery. It covers all the units of the qualification, so you can feel confident that your book fully covers the requirements of your course.

This book contains a number of features to help you acquire the knowledge you need. It also demonstrates the practical skills you will need to master to successfully complete your qualification. We've included additional features to show how the skills and knowledge can be applied to the workforce, as well as tips and advice on how you can improve your chances of gaining employment.

The features include:

* chapter openers which list the learning outcomes you must achieve in each unit

* key terms that provide explanations of important terminology that you will need to know and understand

* Did you know? margin notes to provide key facts that are helpful to your learning

* practical tips to explain facts or skills to remember when undertaking practical tasks

* Reed tips to offer advice about work, building your CV and how to apply the skills and knowledge you have learnt in the workplace

* case studies that are based on real tradespeople who have undertaken apprenticeships and explain why the skills and knowledge you learn with your training provider are useful in the workforce

* practical tasks that provide step-by-step directions and illustrations for a range of projects you may do during your course

* Test yourself multiple choice questions that appear at the end of each unit to give you the chance to revise what you have learnt and to practise your assessment (your tutor will give you the answers to these questions).

Further support for this book can be found at our website, www.planetvocational.com/subjects/build

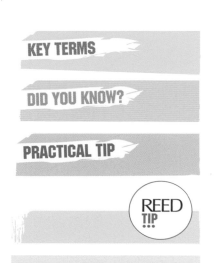

KEY TERMS

DID YOU KNOW?

PRACTICAL TIP

REED TIP

CASE STUDY

PRACTICAL TASK

TEST YOURSELF

Planet Vocational

CONTRIBUTORS TO THIS BOOK

Reed Property & Construction

Reed Property & Construction specialises in placing staff at all levels, in both temporary and permanent positions, across the complete lifecycle of the construction process. Our consultants work with most major construction companies in the UK and our clients are involved with the design, build and maintenance of infrastructure projects throughout the UK.

Expert help

As a leading recruitment consultancy for mid–senior level construction staff in the UK, Reed Property & Construction is ideally placed to advise new workers entering the sector, from building a CV to providing expertise and sharing our extensive sector knowledge with you. That's why, throughout this book, you will find helpful hints from our highly experienced consultants, all designed to help you find that first step on the construction career ladder. These tips range from advice on CV writing to interview tips and techniques, and are all linked in with the learning material in this book.

Work-related advice

Reed Property & Construction has gained insights from some of our biggest clients – leading recruiters within the industry – to help you understand the mind-set of potential employers. This includes the traits and skills that they would like to see in their new employees, why you need the skills taught in this book and how they are used on a day to day basis within their organisations.

Getting your first job

This invaluable information is not available anywhere else and is all geared towards helping you gain a position once you've completed your studies. Entry level positions are not usually offered by recruitment companies, but the advice we've provided will help you to apply for jobs in construction and hopefully gain your first position as a skilled worker.

CONTRIBUTORS TO THIS BOOK

The case studies in this book feature staff from Laing O'Rourke and South Tyneside Homes.

Laing O'Rourke is an international engineering company that constructs large-scale building projects all over the world. Originally formed from two companies, John Laing (founded in 1848) and R O'Rourke and Son (founded in 1978) joined forces in 2001.

At Laing O'Rourke, there is a strong and unique apprenticeship programme. It runs a four-year 'Apprenticeship Plus' scheme in the UK, combining formal college education with on-the-job training. Apprentices receive support and advice from mentors and experienced tradespeople, and are given the option of three different career pathways upon completion remaining on site, continuing into a further education programme, or progressing into supervision and management.

The company prides itself on its people development, supporting educational initiatives and investing in its employees. Laing O'Rourke believes in collaboration and teamwork as a path to achieving greater success, and strives to maintain exceptionally high standards in workplace health and safety.

South Tyneside Council's
Housing Company

South Tyneside Homes was launched in 2006, and was previously part of South Tyneside Council. It now works in partnership with the council to repair and maintain 18,000 properties within the borough, including delivering parts of the Decent Homes Programme.

South Tyneside Homes believes in putting back into the community, with 90 per cent of its employees living in the borough itself. Equality and diversity, as well as health and wellbeing of staff, is a top priority, and it has achieved the Gold Status Investors in People Award.

South Tyneside Homes is committed to the development of its employees, providing opportunities for further education and training and great career paths within the company – 80 per cent of its management team started as apprentices with the company. As well as looking after its staff and their community, the company looks after the environment too, running a renewable energy scheme for council tenants in order to reduce carbon emissions and save tenants money.

The apprenticeship programme at South Tyneside Homes has been recognised nationally, having trained over 80 young people in five main trade areas over the past six years. One of the UK's Top 100 Apprenticeship Employers, it is an Ambassador on the panel of the National Apprentice Service. It has won the Large Employer of the Year Award at the National Apprenticeship Awards and several of its apprentices have been nominated for awards, including winning the Female Apprentice of the Year for the local authority.

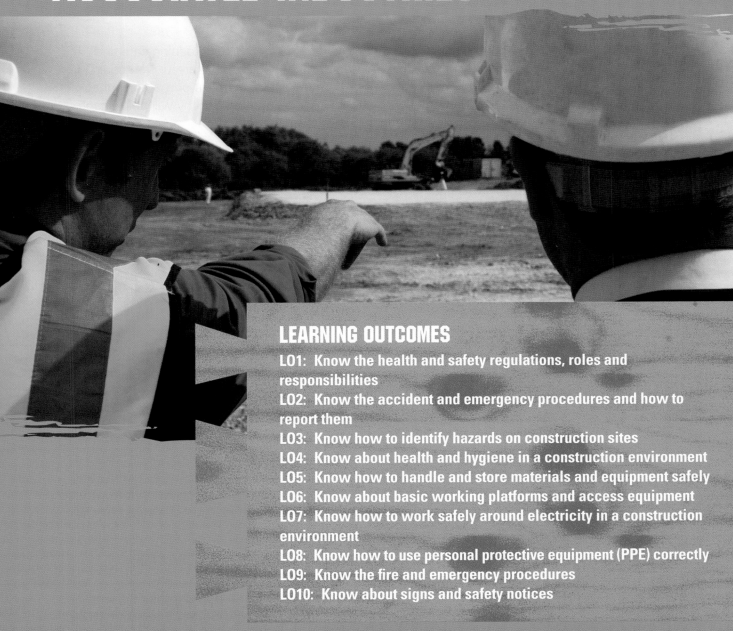

Chapter 1

Unit CSA–L1Core01

HEALTH, SAFETY AND WELFARE IN CONSTRUCTION AND ASSOCIATED INDUSTRIES

LEARNING OUTCOMES

LO1: Know the health and safety regulations, roles and responsibilities

LO2: Know the accident and emergency procedures and how to report them

LO3: Know how to identify hazards on construction sites

LO4: Know about health and hygiene in a construction environment

LO5: Know how to handle and store materials and equipment safely

LO6: Know about basic working platforms and access equipment

LO7: Know how to work safely around electricity in a construction environment

LO8: Know how to use personal protective equipment (PPE) correctly

LO9: Know the fire and emergency procedures

LO10: Know about signs and safety notices

INTRODUCTION

The aim of this chapter is to:

* help you to source relevant safety information
* help you to use the relevant safety procedures at work.

KEY TERMS

HASAWA

– the Health and Safety at Work etc. Act outlines your and your employer's health and safety responsibilities.

COSHH

– the Control of Substances Hazardous to Health Regulations are concerned with controlling exposure to hazardous materials.

DID YOU KNOW?

In 2011 to 2012, there were 49 fatal accidents in the construction industry in the UK. (*Source* HSE, www.hse.gov.uk)

KEY TERMS

HSE

– the Health and Safety Executive, which ensures that health and safety laws are followed.

Accident book

– this is required by law under the Social Security (Claims and Payments) Regulations 1979. Even minor accidents need to be recorded by the employer. For the purposes of RIDDOR, hard copy accident books or online records of incidents are equally acceptable.

HEALTH AND SAFETY REGULATIONS, ROLES AND RESPONSIBILITIES

The construction industry can be dangerous, so keeping safe and healthy at work is very important. If you are not careful, you could injure yourself in an accident or perhaps use equipment or materials that could damage your health. Keeping safe and healthy will help ensure that you have a long and injury-free career.

Although the construction industry is much safer today than in the past, more than 2,000 people are injured and around 50 are killed on site every year. Many others suffer from long-term ill-health such as deafness, spinal damage, skin conditions or breathing problems.

Key health and safety legislation

Laws have been created in the UK to try to ensure safety at work. Ignoring the rules can mean injury or damage to health. It can also mean losing your job or being taken to court.

The two main laws are the Health and Safety at Work etc. Act (**HASAWA**) and the Control of Substances Hazardous to Health Regulations (**COSHH**).

The Health and Safety at Work etc. Act (HASAWA) (1974)
This law applies to all working environments and to all types of worker, sub-contractor, employer and all visitors to the workplace. It places a duty on everyone to follow rules in order to ensure health, safety and welfare. Businesses must manage health and safety risks, for example by providing appropriate training and facilities. The Act also covers first aid, accidents and ill health.

Reporting of Injuries, Diseases and Dangerous Occurrences Regulations (RIDDOR) (1995)
Under RIDDOR, employers are required to report any injuries, diseases or dangerous occurrences to the **Health and Safety Executive (HSE)**. The regulations also state the need to maintain an **accident book**.

Control of Substances Hazardous to Health (COSHH) (2002)

In construction, it is common to be exposed to substances that could cause ill health. For example, you may use oil-based paints or preservatives, or work in conditions where there is dust or bacteria.

Employers need to protect their employees from the risks associated with using hazardous substances. This means assessing the risks and deciding on the necessary precautions to take.

Any control measures (things that are being done to reduce the risk of people being hurt or becoming ill) have to be introduced into the workplace and maintained; this includes monitoring an employee's exposure to harmful substances. The employer will need to carry out health checks and ensure that employees are made aware of the dangers and are supervised.

Control of Asbestos at Work Regulations (2012)

Asbestos was a popular building material in the past because it was a good insulator, had good fire protection properties and also protected metals against corrosion. Any building that was constructed before 2000 is likely to have some asbestos. It can be found in pipe insulation, boilers and ceiling tiles. There is also asbestos cement roof sheeting and there is a small amount of asbestos in decorative coatings such as Artex.

Asbestos has been linked with lung cancer, other damage to the lungs and breathing problems. The regulations require you and your employer to take care when dealing with asbestos:

* You should always assume that materials contain asbestos unless it is obvious that they do not.

* A record of the location and condition of asbestos should be kept.

* A risk assessment should be carried out if there is a chance that anyone will be exposed to asbestos.

The general advice is as follows:

* Do not remove the asbestos. It is not a hazard unless it is removed or damaged.

* Remember that not all asbestos presents the same risk. Asbestos cement is less dangerous than pipe insulation.

* Call in a specialist if you are uncertain.

Provision and Use of Work Equipment Regulations (PUWER) (1998)

PUWER concerns health and safety risks related to equipment used at work. It states that any risks arising from the use of equipment must either be prevented or controlled, and all suitable safety measures must have been taken. In addition, tools need to be:

* suitable for their intended use

* safe

REED TIP

Employers will want to know that you understand the importance of health and safety. Make sure you know the reasons for each safe working practice.

* well maintained

* used only by those who have been trained to do so.

Manual Handling Operations Regulations (1992)

These regulations try to control the risk of injury when lifting or handling bulky or heavy equipment and materials. The regulations state as follows:

* Hazardous manual handling should be avoided if possible.

* An assessment of hazardous manual handling should be made to try to find alternatives.

* You should use mechanical assistance where possible.

* The main idea is to look at how manual handling is carried out and finding safer ways of doing it.

Personal Protection at Work Regulations (PPE) (1992)

This law states that employers must provide employees with personal protective equipment **(PPE)** at work whenever there is a risk to health and safety. PPE needs to be:

* suitable for the work being done

* well maintained and replaced if damaged

* properly stored

* correctly used (which means employees need to be trained in how to use the PPE properly).

Work at Height Regulations (2005)

Whenever a person works at any height there is a risk that they could fall and injure themselves. The regulations place a duty on employers or anyone who controls the work of others. This means that they need to:

* plan and organise the work

* make sure those working at height are **competent**

* assess the risks and provide appropriate equipment

* manage work near or on fragile surfaces

* ensure equipment is inspected and maintained.

In all cases the regulations suggest that, if it is possible, work at height should be avoided. Perhaps the job could be done from ground level? If it is not possible, then equipment and other measures are needed to prevent the risk of falling. When working at height measures also need to be put in place to minimise the distance someone might fall.

KEY TERMS

PPE

– personal protective equipment can include gloves, goggles and hard hats.

Competent

– to be competent an organisation or individual must have:

* sufficient knowledge of the tasks to be undertaken and the risks involved

* the experience and ability to carry out their duties in relation to the project, to recognise their limitations and take appropriate action to prevent harm to those carrying out construction work, or those affected by the work.

(*Source* HSE)

Figure 1.1 Examples of personal protective equipment

Employer responsibilities under HASAWA

HASAWA states that employers with five or more staff need their own health and safety policy. Employers must assess any risks that may be involved in their workplace and then introduce controls to reduce these risks. These risk assessments need to be reviewed regularly.

Employers also need to supply personal protective equipment (PPE) to all employees when it is needed and to ensure that it is worn when required.

Specific employer responsibilities are outlined in Table 1.1.

Employee responsibilities under HASAWA

HASAWA states that all those operating in the workplace must aim to work in a safe way. For example, they must wear any PPE provided and look after their equipment. Employees should not be charged for PPE or any actions that the employer needs to take to ensure safety.

Specific employer responsibilities are outlined in Table 1.1. Table 1.2 identifies the key employee responsibilities.

KEY TERMS

Risk

– the likelihood that a person may be harmed if they are exposed to a hazard.

Hazard

– a potential source of harm, injury or ill-health.

Near miss

– any incident, accident or emergency that did not result in an injury but could have done so.

Employer responsibility	Explanation
Safe working environment	Where possible all potential risks and hazards should be eliminated.
Adequate staff training	When new employees begin a job their induction should cover health and safety. There should be ongoing training for existing employees on risks and control measures.
Health and safety information	Relevant information related to health and safety should be available for employees to read and have their own copies.
Risk assessment	Each task or job should be investigated and potential risks identified so that measures can be put in place. A risk assessment and method statement should be produced. The method statement will tell you how to carry out the task, what PPE to wear, equipment to use and the sequence of its use.
Supervision	A competent and experienced individual should always be available to help ensure that health and safety problems are avoided.

Table 1.1 Employer responsibilities under HASAWA

Employee responsibility	Explanation
Working safely	Employees should take care of themselves, only do work that they are competent to carry out and remove obvious hazards if they are seen.
Working in partnership with the employer	Co-operation is important and you should never interfere with or misuse any health and safety signs or equipment. You should always follow the site rules.
Reporting hazards, near misses and accidents correctly	Any health and safety problems should be reported and discussed, particularly a near miss or an actual accident.

Table 1.2 Employee responsibilities under HASAWA

Health and Safety Executive

The Health and Safety Executive (HSE) is responsible for health, safety and welfare. It carries out spot checks on different workplaces to make sure that the law is being followed.

HSE inspectors have access to all areas of a construction site and can also bring in the police. If they find a problem then they can issue an **improvement notice**. This gives the employer a limited amount of time to put things right.

In serious cases, the HSE can issue a **prohibition notice**. This means all work has to stop until the problem is dealt with. An employer, the employees or **sub-contractors** could be taken to court.

The roles and responsibilities of the HSE are outlined in Table 1.3.

Responsibility	Explanation
Enforcement	It is the HSE's responsibility to reduce work-related death, injury and ill health. It will use the law against those who put others at risk.
Legislation and advice	The HSE will use health and safety legislation to serve improvement or prohibition notices or even to prosecute those who break health and safety rules. Inspectors will provide advice either face-to-face or in writing on health and safety matters.
Inspection	The HSE will look at site conditions, standards and practices and inspect documents to make sure that businesses and individuals are complying with health and safety law.

Table 1.3 HSE roles and responsibilities

Sources of health and safety information

There is a wide variety of health and safety information. Most of it is available free of charge, while other organisations may make a charge to provide information and advice. Table 1.4 outlines the key sources of health and safety information.

Source	Types of information	Website
Health and Safety Executive (HSE)	The HSE is the primary source of work-related health and safety information. It covers all possible topics and industries.	www.hse.gov.uk
Construction Industry Training Board (CITB)	The national training organisation provides key information on legislation and site safety.	www.citb.co.uk
British Standards Institute (BSI)	Provides guidelines for risk management, PPE, fire hazards and many other health and safety-related areas.	www.bsigroup.com
Royal Society for the Prevention of Accidents (RoSPA)	Provides training, consultancy and advice on a wide range of health and safety issues that are aimed to reduce work related accidents and ill health.	www.rospa.com
Royal Society for Public Health (RSPH)	Has a range of qualifications and training programmes focusing on health and safety.	www.rsph.org.uk

Table 1.4 Health and safety information

Informing the HSE

The HSE requires the reporting of:

* deaths and injuries – any **major injury**, **over 7-day injury** or death

* occupational disease

* dangerous occurrence – a collapse, explosion, fire or collision

* gas accidents – any accidental leaks or other incident related to gas.

Enforcing guidance

Work-related injuries and illnesses affect huge numbers of people. According to the HSE, 1.1 million working people in the UK suffered from a work-related illness in 2011 to 2012. Across all industries, 173 workers were killed, 111,000 other injuries were reported and 27 million working days were lost.

The construction industry is a high risk one and, although only around 5 per cent of the working population is in construction, it accounts for 10 per cent of all major injuries and 22 per cent of fatal injuries.

The good news is that enforcing guidance on health and safety has driven down the numbers of injuries and deaths in the industry. Only 20 years ago over 120 construction workers died in workplace accidents each year. This is now reduced to fewer than 60 a year.

However, there is still more work to be done and it is vital that organisations such as the HSE continue to enforce health and safety and continue to reduce risks in the industry.

On-site safety inductions and toolbox talks

The HSE suggests that all new workers arriving on site should attend a short induction session on health and safety. It should:

* show the commitment of the company to health and safety

* explain the health and safety policy

* explain the roles individuals play in the policy

* state that each individual has a legal duty to contribute to safe working

* cover issues like excavations, work at height, electricity and fire risk

* provide a layout of the site and show evacuation routes

* identify where fire fighting equipment is located

* ensure that all employees have evidence of their skills

* stress the importance of signing in and out of the site.

KEY TERMS

Major injury

– any fractures, amputations, dislocations, loss of sight or other severe injury.

Over 7-day injury

– an injury that has kept someone off work for more than seven days.

DID YOU KNOW?

Workplace injuries cost the UK £13.4bn in 2010 to 2011.

DID YOU KNOW?

Toolbox talks are normally given by a supervisor and often take place on site, either during the course of a normal working day or when someone has been seen working in an unsafe way. CITB produces a book called *GT700 Toolbox Talks* which covers a range of health and safety topics, from trying a new process and using new equipment to particular hazards or work practices.

Behaviour and actions that could affect others

It is the responsibility of everyone on site not only to look after their own health and safety, but also to ensure that their actions do not put anyone else at risk.

Trying to carry out work that you are not competent to do is not only dangerous to yourself but could compromise the safety of others.

Simple actions, such as ensuring that all of your rubbish and waste is properly disposed of, will go a long way to removing hazards on site that could affect others.

Just as you should not create a hazard, ignoring an obvious one is just as dangerous. You should always obey site rules and particularly the health and safety rules. You should follow any instructions you are given.

ACCIDENT AND EMERGENCY PROCEDURES

PRACTICAL TIP

If you come across any health and safety problems you should report them so that they can be controlled.

All sites will have specific procedures for dealing with accidents and emergencies. An emergency will often mean that the site needs to be evacuated, so you should know in advance where to assemble and who to report to. The site should never be re-entered without authorisation from an individual in charge or the emergency services.

Types of emergencies

Emergencies are incidents that require immediate action. They can include:

* fires
* spillages or leaks of chemicals or other hazardous substances, such as gas
* failure of a scaffold
* collapse of a wall or trench
* a health problem
* an injury
* bombs and security alerts.

Legislation and reporting accidents

RIDDOR (1995) puts a duty on employers, anyone who is self-employed, or an individual in control of the work, to report any serious workplace accidents, occupational diseases or dangerous occurrences (also known as near misses).

The report has to be made by these individuals and, if it is serious enough, the responsible person may have to fill out a RIDDOR report.

Figure 1.2 It's important that you know where your company's fire-fighting equipment is located

Injuries, diseases and dangerous occurrences

Construction sites can be dangerous places, as we have seen. The HSE maintains a list of all possible injuries, diseases and dangerous occurrences, particularly those that need to be reported.

Injuries

There are two main classifications of injuries: minor and major. A minor injury can usually be handled by a competent first aider, although it is often a good idea to refer the individual to their doctor or to the hospital. Typical minor injuries can include:

- minor cuts
- minor burns
- exposure to fumes.

Major injuries are more dangerous and will usually require the presence of an ambulance with paramedics. Major injuries can include:

- bone fracture
- concussion
- unconsciousness
- electric shock.

Diseases

There are several different diseases and health issues that have to be reported, particularly if a doctor notifies that a disease has been diagnosed. These include:

- poisoning
- infections
- skin diseases
- occupational cancer
- lung diseases
- hand/arm vibration syndrome.

Dangerous occurrences

Even if something happens that does not result in an injury, but could easily have done so, it is classed as a dangerous occurrence. It needs to be reported immediately and then followed up by an accident report form. Dangerous occurrences can include:

- accidental release of a substance that could damage health

- anything coming into contact with overhead power lines

- an electrical problem that caused a fire or explosion

- collapse or partial collapse of scaffolding over 5 m high.

PRACTICAL TIP

An up-to-date list of dangerous occurrences is maintained by the Health and Safety Executive.

Recording accidents and emergencies

The Reporting of Injuries, Diseases and Dangerous Occurrences Regulations (RIDDOR) (1995) requires employers to:

- report any relevant injuries, diseases or dangerous occurrences to the Health and Safety Executive (HSE)

- keep records of incidents in a formal and organised manner (for example, in an accident book or online database).

After an accident, you may need to complete an accident report form – either in writing or online. This form may be completed by the person who was injured or the first aider.

On the accident report form you need to note down:

* the casualty's personal details, e.g. name, address, occupation

* the name of the person filling in the report form

* the details of the accident.

In addition, the person reporting the accident will need to sign the form.

On site a trained first aider will be the first individual to try and deal with the situation. In addition to trying to save life, stop the condition from getting worse and getting help, they will also record the occurrence.

On larger sites there will be a health and safety officer, who would keep records and documentation detailing any accidents and emergencies that have taken place on site. All companies should keep such records; it may be a legal requirement for them to do so under RIDDOR and it is good practice to do so in case the HSE asks to see it.

Importance of reporting accidents and near misses

Reporting incidents is not just about complying with the law or providing information for statistics. Each time an accident or near miss takes place it means lessons can be learned and future problems avoided.

The accident or near miss can alert the business or organisation to a potential problem. They can then take steps to ensure that it does not occur in the future.

Major and minor injuries and near misses

RIDDOR defines a major injury as:

* a fracture (but not to a finger, thumb or toes)

* a dislocation

* an amputation

* a loss of sight in an eye

* a chemical or hot metal burn to the eye

* a penetrating injury to the eye

* an electric shock or electric burn leading to unconsciousness and/or requiring resuscitation

* hyperthermia, heat-induced illness or unconsciousness

* asphyxia

* exposure to a harmful substance

* inhalation of a substance

* acute illness after exposure to toxins or infected materials.

A minor injury could be considered as any occurrence that does not fall into any of the above categories.

A near miss is any incident that did not actually result in an injury but which could have caused a major injury if it had done so. Non-reportable near misses are useful to record as they can help to identify potential problems. Looking at a list of near misses might show patterns for potential risk.

Accident trends

We have already seen that the HSE maintains statistics on the number and types of construction accidents. The following are among the 2011/2012 construction statistics:

* There were 49 fatalities.

* There were 5,000 occupational cancer patients.

* There were 74,000 cases of work-related ill health.

* The most common types of injury were caused by falls, although many injuries were caused by falling objects, collapses and electricity. A number of construction workers were also hurt when they slipped or tripped, or were injured while lifting heavy objects.

Accidents, emergencies and the employer

Even less serious accidents and injuries can cost a business a great deal of money. But there are other costs too:

* Poor company image – if a business does not have health and safety controls in place then it may get a reputation for not caring about its employees. The number of accidents and injuries may be far higher than average.

* Loss of production – the injured individual might have to be treated and then may need a period of time off work to recover. The loss of production can include those who have to take time out from working to help the injured person and the time of a manager or supervisor who has to deal with all the paperwork and problems.

* Insurance – each time there is an accident or injury claim against the company's insurance the premiums will go up. If there are many accidents and injuries the business may find it impossible to get insurance. It is a legal requirement for a business to have insurance so in the end that company might have to close down.

* Closure of site – if there is a serious accident or injury then the site may have to be closed while investigations take place to discover the reason, or who was responsible. This could cause serious delays and loss of income for workers and the business.

DID YOU KNOW?

RoSPA (the Royal Society for the Prevention of Accidents) uses many of the statistics from the HSE. The latest figures that RoSPA has analysed date back to 2008/2009. In that year, 1.2 million people in the UK were suffering from work-related illnesses. With fewer than 132,000 reportable injuries at work, this is believed to be around half of the real figure.

DID YOU KNOW?

An employee working in a small business broke two bones in his arm. He could not return to proper duties for eight months. He lost out on wages while he was off sick and, in total, it cost the business over £45,000.

REED TIP

On some construction sites, you may get a Health and Safety Inspector come to look round without any notice – one more reason to always be thinking about working safely.

Accident and emergency authorised personnel

Several different groups of people could be involved in dealing with accident and emergency situations. These are listed in Table 1.5.

Authorised personnel	Role
First aiders and emergency responders	These are employees on site and in the workforce who have been trained to be the first to respond to accidents and injuries. The minimum provision of an appointed person would be someone who has had basic first aid training. The appointment of a first aider is someone who has attained a higher or specific level of training. A construction site with fewer than 5 employees needs an appointed first aider. A construction site with up to 50 employees requires a trained first aider, and for bigger sites at least one trained first aider is required for every 50 people.
Supervisors and managers	These have the responsibility of managing the site and would have to organise the response and contact emergency services if necessary. They would also ensure that records of any accidents are completed and up to date and notify the HSE if required.
Health and Safety Executive	The HSE requires businesses to investigate all accidents and emergencies. The HSE may send an inspector, or even a team, to investigate and take action if the law has been broken.
Emergency services	Calling the emergency services depends on the seriousness of the accident. Paramedics will take charge of the situation if there is a serious injury and if they feel it necessary will take the individual to hospital.

Table 1.5 People who deal with accident and emergency situations

Figure 1.3 A typical first aid box

The basic first aid kit

BS 8599 relates to first aid kits, but it is not legally binding. The contents of a first aid box will depend on an employer's assessment of their likely needs. The HSE does not have to approve the contents of a first aid box but it states that where the work involves low level hazards the minimum contents of a first aid box should be:

* a copy of its leaflet on first aid – *HSE Basic advice on first aid at work*
* 20 sterile plasters of assorted size
* 2 sterile eye pads
* 4 sterile triangular bandages
* 6 safety pins
* 2 large sterile, unmedicated wound dressings
* 6 medium-sized sterile unmedicated wound dressings
* 1 pair of disposable gloves.

The HSE also recommends that no tablets or medicines are kept in the first aid box.

What to do if you discover an accident

When an accident happens it may not only injure the person involved directly, but it may also create a hazard that could then injure others. You need to make sure that the area is safe enough for you or someone else to help the injured person. It may be necessary to turn off the electrical supply or remove obstructions to the site of the accident.

The first thing that needs to be done if there is an accident is to raise the alarm. This could mean:

* calling for the first aider

* phoning for the emergency services

* dealing with the problem yourself.

How you respond will depend on the severity of the injury.

You should follow this procedure if you need to contact the emergency services:

* Find a telephone away from the emergency.

* Dial 999.

* You may have to go through a switchboard. Carefully listen to what the operator is saying to you and try to stay calm.

* When asked, give the operator your name and location, and the name of the emergency service or services you require.

* You will then be transferred to the appropriate emergency service, who will ask you questions about the accident and its location. Answer the questions in a clear and calm way.

* Once the call is over, make sure someone is available to help direct the emergency services to the location of the accident.

IDENTIFYING HAZARDS

As we have already seen, construction sites are potentially dangerous places. The most effective way of handling health and safety on a construction site is to spot the hazards and deal with them before they can cause an accident or an injury. This begins with basic housekeeping and carrying out risk assessments. It also means having a procedure in place to report hazards so that they can be dealt with.

Good housekeeping

Work areas should always be clean and tidy. Sites that are messy, strewn with materials, equipment, wires and other hazards can prove to be very dangerous. You should:

* always work in a tidy way

* never block fire exits or emergency escape routes

* never leave nails and screws scattered around

* ensure you clean and sweep up at the end of each working day

* not block walkways

* never overfill skips or bins

* never leave food waste on site.

Risk assessments and method statements

It is a legal requirement for employers to carry out risk assessments. This covers not only those who are actually working on a particular job, but other workers in the immediate area, and others who might be affected by the work.

It is important to remember that when you are carrying out work your actions may affect the safety of other people. It is important, therefore, to know whether there are any potential hazards. Once you know what these hazards are you can do something to either prevent or reduce them as a risk. Every job has potential hazards.

There are five simple steps to carrying out a risk assessment, which are shown in Table 1.6, using the example of repointing brickwork on the front face of a dwelling.

Step	Action	Example
1	Identify hazards	The property is on a street with a narrow pavement. The damaged brickwork and loose mortar need to be removed and placed in a skip below. Scaffolding has been erected. The road is not closed to traffic.
2	Identify who is at risk	The workers repointing are at risk as they are working at height. Pedestrians and vehicles passing are at risk from the positioning of the skip and the chance that debris could fall from height.
3	What is the risk from the hazard that may cause an accident?	The risk to the workers is relatively low as they have PPE and the scaffolding has been correctly erected. The risk to those passing by is higher, as they are unaware of the work being carried out above them.
4	Measures to be taken to reduce the risk	Station someone near the skip to direct pedestrians and vehicles away from the skip while the work is being carried out. Fix a secure barrier to the edge of the scaffolding to reduce the chance of debris falling down. Lower the bricks and mortar debris using a bucket or bag into the skip and not throwing them from the scaffolding. Consider carrying out the work when there are fewer pedestrians and less traffic on the road.
5	Monitor the risk	If there are problems with the first stages of the job, you need to take steps to solve them. If necessary consider taking the debris by hand through the building after removal.

Table 1.6 A five-step risk assessment for repointing brickwork

Your employer should follow these working practices, which can help to prevent accidents or dangerous situations occurring in the workplace:

* *Risk assessments* look carefully at what could cause an individual harm and how to prevent this. This is to ensure that no one should be injured or become ill as a result of their work. Risk assessments identify how likely it is that an accident might happen and the consequences of it happening. A risk factor is worked out and control measures created to try to offset them.

* *Method statements,* however brief, should be available for every risk assessment. They summarise risk assessments and other findings to provide guidance on how the work should be carried out.

* *Permit to work systems* are used for very high risk or even potentially fatal activities. They are checklists that need to be completed before the work begins. They must be signed by a supervisor.

* *A hazard book* lists standard tasks and identifies common hazards. These are useful tools to help quickly identify hazards related to particular tasks.

Types of hazards

Typical construction accidents can include:

* fires and explosions

* slips, trips and falls.

* burns, including those from chemicals

* falls from scaffolding, ladders and roofs

* electrocution

* injury from faulty machinery

* power tool accidents

* being hit by construction debris

* falling through holes in flooring

We will look at some of the more common hazards in a little more detail.

Fires
Fires need oxygen, heat and fuel to burn. Even a spark can provide enough heat needed to start a fire, and anything flammable, such as petrol, paper or wood, provides the fuel. It may help to remember the 'triangle of fire' – heat, oxygen and fuel are all needed to make fire so remove one or more to help prevent or stop the fire.

Tripping

Leaving equipment and materials lying around can cause accidents, as can trailing cables and spilt water or oil. Some of these materials are also potential fire hazards.

Chemical spills

If the chemicals are not hazardous then they just need to be mopped up. But sometimes they do involve hazardous materials and there will be an existing plan on how to deal with them. A risk assessment will have been carried out.

Falls from height

A fall even from a low height can cause serious injuries. Precautions need to be taken when working at height to avoid permanent injury. You should also consider falls into open excavations as falls from height. All the same precautions need to be in place to prevent a fall.

Burns

Burns can be caused not only by fires and heat, but also from chemicals and solvents. Electricity and wet concrete and cement can also burn skin. PPE is often the best way to avoid these dangers. Sunburn is a common and uncomfortable form of burning and sunscreen should be made available. For example, keeping skin covered up will help to prevent sunburn. You might think a tan looks good, but it could lead to skin cancer.

Electrical

Electricity is hazardous and electric shocks can cause burns and muscle damage, and can kill.

Exposure to hazardous substances

We look at hazardous substances in more detail on pages 20–1. COSHH regulations identify hazardous substances and require them to be labelled. You should always follow the instructions when using them.

Plant and vehicles

On busy sites there is always a danger from moving vehicles and heavy plant. Although many are fitted with reversing alarms, it may not be easy to hear them over other machinery and equipment. You should always ensure you are not blocking routes or exits. Designated walkways separate site traffic and pedestrians – this includes workers who are walking around the site. Crossing points should be in place for ease of movement on site.

Reporting hazards

We have already seen that hazards have the potential to cause serious accidents and injuries. It is therefore important to report hazards and there are different methods of doing this.

The first major reason to report hazards is to prevent danger to others, whether they are other employees or visitors to the site. It is vital to prevent accidents from taking place and to quickly correct any dangerous situations.

Injuries, diseases and actual accidents all need to be reported and so do dangerous occurrences. These are incidents that do not result in an actual injury, but could easily have hurt someone.

Accidents need to be recorded in an accident book, computer database or other secure recording system, as do near misses. Again it is a legal requirement to keep appropriate records of accidents and every company will have a procedure for this which they should tell you about. Everyone should know where the book is kept or how the records are made. Anyone that has been hurt or has taken part in dealing with an occurrence should complete the details of what has happened. Typically this will require you to fill in:

* the date, time and place of the incident

* how it happened

* what was the cause

* how it was dealt with

* who was involved

* signature and date.

The details in the book have to be transferred onto an official HSE report form.

As far as is possible, the site, company or workplace will have set procedures in place for reporting hazards and accidents. These procedures will usually be found in the place where the accident book or records are stored. The location tends to be posted on the site notice board.

How hazards are created

Construction sites are busy places. There are constantly new stages in development. As each stage is begun a whole new set of potential hazards need to be considered.

At the same time, new workers will always be joining the site. It is mandatory for them to be given health and safety instruction during induction. But sometimes this is impossible due to pressure of work or availability of trainers.

Construction sites can become even more hazardous in times of extreme weather:

* Flooding – long periods of rain can cause trenches to fill with water, cellars to be flooded and smooth surfaces to become extremely wet and slippery.

* Wind – strong winds may prevent all work at height. Scaffolding may have become unstable, unsecured roofing materials may come loose, dry-stored materials such as sand and cement may have been blown across the site.

* Heat – this can change the behaviour of materials: setting quicker, failing to cure and melting. It can also seriously affect the health of the workforce through dehydration and heat exhaustion.

* Snow – this can add enormous weight to roofs and other structures and could cause collapse. Snow can also prevent access or block exits and can mean that simple and routine work becomes impossible due to frozen conditions.

Storing combustibles and chemicals

A combustible substance can be both flammable and explosive. There are some basic suggestions from the HSE about storing these:

* Ventilation – the area should be well ventilated to disperse any vapours that could trigger off an explosion.

* Ignition – an ignition is any spark or flame that could trigger off the vapours, so materials should be stored away from any area that uses electrical equipment or any tool that heats up.

* Containment – the materials should always be kept in proper containers with lids and there should be spillage trays to prevent any leak seeping into other parts of the site.

* Exchange – in many cases it can be possible to find an alternative material that is less dangerous. This option should be taken if possible.

* Separation – always keep flammable substances away from general work areas. If possible they should be partitioned off.

Combustible materials can include a large number of commonly used substances, such as cleaning agents, paints and adhesives.

HEALTH AND HYGIENE

Just as hazards can be a major problem on site, other less obvious problems relating to health and hygiene can also be an issue. It is both your responsibility and that of your employer to make sure that you stay healthy.

The employer will need to provide basic welfare facilities, no matter where you are working and these must have minimum standards.

Welfare facilities

Welfare facilities can include a wide range of different considerations, as can be seen in Table 1.7

DID YOU KNOW?

You do not have to be involved in specialist work to come into contact with combustibles.

KEY TERMS

Contamination

– this is when a substance has been polluted by some harmful substance or chemical.

Facilities	Purpose and minimum standards
Toilets	If there is a lock on the door there is no need to have separate male and female toilets. There should be enough for the site workforce. If there is no flushing water on site they must be chemical toilets.
Washing facilities	There should be a wash basin large enough to be able to wash up to the elbow. There should be soap, hot and cold water and, if you are working with dangerous substances, then showers are needed.
Drinking water	Clean drinking water should be available; either directly connected to the mains or bottled water. Employers must ensure that there is no contamination.
Dry room	This can operate also as a store room, which needs to be secure so that workers can leave their belongings there and also use it as a place to dry out if they have been working in wet weather, in which case a heater needs to be provided.
Work break area	This is a shelter out of the wind and rain, with a kettle, a microwave, tables and chairs. It should also have heating.

Table 1.7 Welfare facilities in the workplace

CASE STUDY

South Tyneside Homes

South Tyneside Council's Housing Company

Staying safe on site

Johnny McErlane finished his apprenticeship at South Tyneside Homes a year ago.

'I've been working on sheltered accommodation for the last year, so there are a lot of vulnerable and elderly people around. All the things I learnt at college from doing the health and safety exams comes into practice really, like taking care when using extension leads, wearing high-vis and correct footwear. It's not just about your health and safety, but looking out for others as well.

On the shelters, you can get a health and safety inspector who just comes around randomly, so you have to always be ready. It just becomes a habit once it's been drilled into you. You're health and safety conscious all the time.

The shelters also have a fire alarm drill every second Monday, so you've got to know the procedure involved there. When it comes to the more specialised skills, such as mouth-to-mouth and CPR, you might have a designated first aider on site who will have their skills refreshed regularly. Having a full first aid certificate would be valuable if you're working in construction.

You cover quite a bit of the first aid skills in college and you really have to know them because you're not always working on large sites. For example, you might be on the repairs team, working in people's houses where you wouldn't have a first aider, so you've got to have the basic knowledge yourself, just in case. All our vans have a basic first aid kit that's kept fully stocked.

The company keeps our knowledge current with these "toolbox talks", which are like refresher courses. They give you any new information that needs to be passed on to all the trades. It's a good way of keeping everyone up to date.'

Noise

Ear defenders are the best precaution to protect the ears from loud noises on site. Ear defenders are either basic ear plugs or ear muffs, which can be seen in Fig 1.13 on page 32.

The long-term impact of noise depends on the intensity and duration of the noise. Basically, the louder and longer the noise exposure, the more damage is caused. There are ways of dealing with this:

* Remove the source of the noise.

* Move the equipment away from those not directly working with it.

* Put the source of the noise into a soundproof area or cover it with soundproof material.

* Ask a supervisor if they can move all other employees away from that part of the site until the noise stops.

Substances hazardous to health

COSHH Regulations (see page 3) identify a wide variety of substances and materials that must be labelled in different ways.

Controlling the use of these substances is always difficult. Ideally, their use should be eliminated (stopped) or they should be replaced with something less harmful. Failing this, they should only be used in controlled or restricted areas. If none of this is possible then they should only be used in controlled situations.

If a hazardous situation occurs at work, then you should:

* ensure the area is made safe

* inform the supervisor, site manager, safety officer or other nominated person.

You will also need to report any potential hazards or near misses.

Personal hygiene

Construction sites can be dirty places to work. Some jobs will expose you to dust, chemicals or substances that can make contact with your skin or may stain your work clothing. It is good practice to wear suitable PPE as a first line of defence as chemicals can penetrate your skin. Whenever you have finished a job you should always wash your hands. This is certainly true before eating lunch or travelling home. It can be good practice to have dedicated work clothing, which should be washed regularly.

Always ensure you wash your hands and face and scrub your nails. This will prevent dirt, chemicals and other substances from contaminating your food and your home.

Make sure that you regularly wash your work clothing and either repair it or replace it if it becomes too worn or stained.

Health risks

The construction industry uses a wide variety of substances that could harm your health. You will also be carrying out work that could be a health risk to you, and you should always be aware that certain activities could cause long-term damage or even kill you if things go wrong. Unfortunately not all health risks are immediately obvious. It is important to make sure that from time to time you have health checks, particularly if you have been using hazardous substances. Table 1.8 outlines some potential health risks in a typical construction site.

KEY TERMS

Dermatitis

– this is an inflammation of the skin. The skin will become red and sore, particularly if you scratch the area. A GP should be consulted.

Leptospirosis

– this is also known as Weil's disease. It is spread by touching soil or water contaminated with the urine of wild animals infected with the leptospira bacteria. Symptoms are usually flu-like but in extreme cases it can cause organ failure.

Health risk	Potential future problems
Dust	The most dangerous potential dust is, of course, asbestos, which **should only be handled by specialists under controlled conditions**. But even brick dust and other fine particles can cause eye injuries, problems with breathing and even cancer.
Chemicals	Inhaling or swallowing dangerous chemicals could cause immediate, long-term damage to lungs and other internal organs. Skin problems include burns or skin can become very inflamed and sore. This is known as dermatitis.
Bacteria	Contact with waste water or soil could lead to a bacterial infection. The germs in the water or dirt could cause infection which will require treatment if they enter the body. The most extreme version is leptospirosis.
Heavy objects	Lifting heavy, bulky or awkward objects can lead to permanent back injuries that could require surgery. Heavy objects can also damage the muscles in all areas of the body.
Noise	Failure to wear ear defenders when you are exposed to loud noises can permanently affect your hearing. This could lead to deafness in the future.
Vibrating tools	Using machines that vibrate can cause a condition known as hand/arm vibration syndrome (HAVS) or vibration white finger, which is caused by injury to nerves and blood vessels. You will feel tingling that could lead to permanent numbness in the fingers and hands, as well as muscle weakness.
Cuts	Any open wound, no matter how small, leaves your body exposed to potential infections. Cuts should always be cleaned and covered, preferably with a waterproof dressing. The blood loss from deep cuts could make you feel faint and weak, which may be dangerous if you are working at height or operating machinery.
Sunlight	Most construction work involves working outside. There is a temptation to take advantage of hot weather and get a tan. But long-term exposure to sunshine means risking skin cancer so you should cover up and apply sun cream.
Head injuries	You should seek medical attention after any bump to the head. Severe head injuries could cause epilepsy, hearing problems, brain damage or death.

Table 1.8 Health risks in construction

HANDLING AND STORING MATERIALS AND EQUIPMENT

On a busy construction site it is often tempting not to even think about the potential dangers of handling equipment and materials. If something needs to be moved or collected you will just pick it up without any thought. It is also tempting just to drop your tools and other equipment when you have finished with them to deal with later. But abandoned equipment and tools can cause hazards both for you and for other people.

Safe lifting

Lifting or handling heavy or bulky items is a major cause of injuries on construction sites. So whenever you are dealing with a heavy load, it is important to carry out a basic risk assessment.

The first thing you need to do is to think about the job to be done and ask:

* Do I need to lift it manually or is there another way of getting the object to where I need it?

Consider any mechanical methods of transporting loads or picking up materials. If there really is no alternative, then ask yourself:

1. Do I need to bend or twist?

2. Does the object need to be lifted or put down from high up?

3. Does the object need to be carried a long way?

4. Does the object need to be pushed or pulled for a long distance?

5. Is the object likely to shift around while it is being moved?

If the answer to any of these questions is 'yes', you may need to adjust the way the task is done to make it safer.

Think about the object itself. Ask:

1. Is it just heavy or is it also bulky and an awkward shape?

2. How easy is it to get a good hand-hold on the object?

3. Is the object a single item or are there parts that might move around and shift the weight?

4. Is the object hot or does it have sharp edges?

Again, if you have answered 'yes' to any of these questions, then you need to take steps to address these issues.

It is also important to think about the working environment and where the lifting and carrying is taking place. Ask yourself:

1. Are the floors stable?

2. Are the surfaces slippery?

3. Will a lack of space restrict my movement?

4. Are there any steps or slopes?

5. What is the lighting like?

Before lifting and moving an object, think about the following:

* Check that your pathway is clear to where the load needs to be taken.

* Look at the product data sheet and assess the weight. If you think the object is too heavy or difficult to move then ask someone to help you. Alternatively, you may need to use a mechanical lifting device.

When you are ready to lift, gently raise the load. Take care to ensure the correct posture – you should have a straight back, with your elbows tucked in, your knees bent and your feet slightly apart.

Once you have picked up the load, move slowly towards your destination. When you get there, make sure that you do not drop the load but carefully place it down.

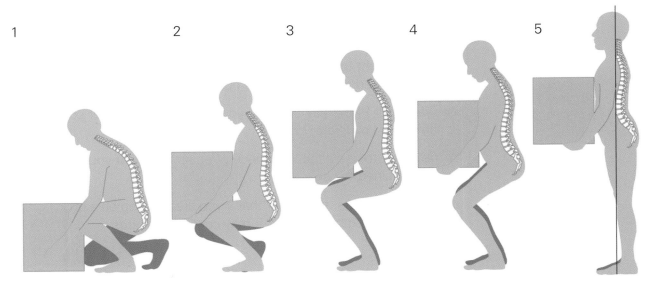

1   2   3   4   5

Figure 1.4 Take care to follow the correct procedure for lifting

Sack trolleys are useful for moving heavy and bulky items around. Gently slide the bottom of the sack trolley under the object and then raise the trolley to an angle of 45° before moving off. Make sure that the object is properly balanced and is not too big for the trolley.

Trailers and forklift trucks are often used on large construction sites, as are dump trucks. Never use these without proper training.

Figure 1.5 Pallet truck

Figure 1.6 Sack trolley

Site safety equipment

You should always read the construction site safety rules and when required wear your PPE. Simple things, such as wearing the right footwear for the right job, are important.

Safety equipment falls into two main categories:

* PPE – including hard hats, footwear, gloves, glasses and safety vests

* perimeter safety – this includes screens, netting and guards or clamps to prevent materials from falling or spreading.

Construction safety is also directed by signs, which will highlight potential hazards.

Safe handling of materials and equipment

All tools and equipment are potentially dangerous. It is up to you to make sure that they do not cause harm to yourself or others. You should always know how to use tools and equipment. This means either instruction from someone else who is experienced, or at least reading the manufacturer's instructions.

You should always make sure that you:

* use the right tool – don't be tempted to use a tool that is close to hand instead of the one that is right for the job

* wear your PPE – the one time you decide not to bother could be the time that you injure yourself

* never try to use a tool or a piece of equipment that you have not been trained to use.

You should always remember that if you are working on a building that was constructed before 2000 it may contain asbestos.

Correct storage

We have already seen that tools and equipment need to be treated with respect. Damaged tools and equipment are not only less effective at doing their job, they could also cause you to injure yourself.

Table 1.9 provides some pointers on how to store and handle different types of materials and equipment.

Materials and equipment	Safe storage and handling
Hand tools	Store hand tools with sharp edges either in a cover or a roll. They should be stored in bags or boxes. They should always be dried before putting them away as they will rust.
Power tools	Never carry them by the cable. Store them in their original carrying case. Always follow the manufacturer's instructions.
Wheelbarrows	Check the tyres and metal stays regularly. Always clean out after use and never overload.
Bricks and blocks	Never store more than two packs high. When cutting open a pack, be careful as the bricks could collapse.
Slabs and curbs	Store slabs flat on their edges on level ground, preferably with wood underneath to prevent damage. Store curbs the same way. To prevent weather damage, cover them with a sheet.
Tiles	Always cover them and protect them from damage as they are relatively fragile. Ideally store them in a hut or container.
Aggregates	Never store aggregates under trees as leaves will drop on them and contaminate them. Cover them with plastic sheets.
Plaster and plasterboard	Plaster needs to be kept dry, so even if stored inside you should take the precaution of putting the bags on pallets. To prevent moisture do not store against walls and do not pile higher than five bags. Plasterboard can be awkward to manage and move around. It also needs to be stored in a waterproof area. It should be stored flat and off the ground but should not be stored against walls as it may bend. Use a rotation system so that the materials are not stored in the same place for long periods.
Wood	Always keep wood in dry, well-ventilated conditions. If it needs to be stored outside it should be stored on bearers that may be on concrete. If wood gets wet and bends it is virtually useless. Always be careful when moving large cuts of wood or sheets of ply or MDF as they can easily become damaged.
Adhesives and paint	Always read the manufacturer's instructions. Ideally they should always be stored on clearly marked shelves. Make sure you rotate the stock using the older stock first. Always make sure that containers are tightly sealed. Storage areas must comply with fire regulations and display signs to advise of their contents.

Table 1.9 Safe storing and handling of materials and equipment

Waste control

The expectation within the building services industry is increasingly that working practices conserve energy and protect the environment. Everyone can play a part in this. For example, you can contribute by turning off hose pipes when you have finished using water, or not running electrical items when you don't need to.

Simple things, such as keeping construction sites neat and orderly, can go a long way to conserving energy and protecting the environment. A good way to remember this is Sort, Set, Shine, Standardise:

* Sort – sort and store items in your work area, eliminate clutter and manage deliveries.

* Set – everything should have its own place and be clearly marked and easy to access. In other words, be neat!

Figure 1.7 It's important to create as little waste as possible on the construction site

* Shine – clean your work area and you will be able to see potential problems far more easily.

* Standardise – by using standardised working practices you can keep organised, clean and safe.

Reducing waste is all about good working practice. By reducing wastage disposal, and recycling materials on site, you will benefit from savings on raw materials and lower transportation costs.

Planning ahead, and accurately measuring and cutting materials, means that you will be able to reduce wastage.

BASIC WORKING PLATFORMS AND ACCESS EQUIPMENT

Working at height should be eliminated or the work carried out using other methods where possible. However, there may be situations where you may need to work at height. These situations can include:

* roofing

* repair and maintenance above ground level

* working on high ceilings.

Any work at height must be carefully planned. Access equipment includes all types of ladder, scaffold and platform. You must always use a working platform that is safe. Sometimes a simple step ladder will be sufficient, but at other times you may have to use a tower scaffold.

Generally, ladders are fine for small, quick jobs of less than 30 minutes. However, for larger, longer jobs a more permanent piece of access equipment will be necessary.

Working platforms and access equipment: good practice and dangers of working at height

Table 1.10 outlines the common types of equipment used to allow you to work at heights, along with the basic safety checks necessary.

Equipment	Main features	Safety checks
Step ladder	Ideal for confined spaces. Four legs give stability	• Knee should remain below top of steps • Check hinges, cords or ropes • Position only to face work
Ladder	Ideal for basic access, short-term work. Made from aluminium, fibreglass or wood	• Check rungs, tie rods, repairs, and ropes and cords on stepladders • Ensure it is placed on firm, level ground • Angle should be no greater than 75° or 1 in 4
Mobile mini towers or scaffolds	These are usually aluminium and foldable, with lockable wheels	• Ensure the ground is even and the wheels are locked • Never move the platform while it has tools, equipment or people on it
Roof ladders and crawling boards	The roof ladder allows access while crawling boards provide a safe passage over tiles	• The ladder needs to be long enough and supported • Check boards are in good condition • Check the welds are intact • Ensure all clips function correctly
Mobile tower scaffolds	These larger versions of mini towers usually have edge protection	• Ensure the ground is even and the wheels are locked • Never move the platform while it has tools, equipment or people on it • Base width to height ratio should be no greater than 1:3
Fixed scaffolds and edge protection	Scaffolds fitted and sized to the specific job, with edge protection and guard rails	• There needs to be sufficient braces, guard rails and scaffold boards • The tubes should be level • There should be proper access using a ladder
Mobile elevated work platforms	Known as scissor lifts or cherry pickers	• Specialist training is required before use • Use guard rails and toe boards • Care needs to be taken to avoid overhead hazards such as cables

Table 1.10 Equipment for working at height and safety checks

You must be trained in the use of certain types of access equipment, like mobile scaffolds. Care needs to be taken when assembling and using access equipment. These are all examples of good practice:

* Step ladders should always rest firmly on the ground. Only use the top step if the ladder is part of a platform.

* Do not rest ladders against fragile surfaces, and always use both hands to climb. It is best if the ladder is steadied (footed) by someone at the foot of the ladder. Always maintain three points of contact – two feet and one hand.

* A roof ladder is positioned by turning it on its wheels and pushing it up the roof. It then hooks over the ridge tiles. Ensure that the access ladder to the roof is directly beside the roof ladder.

* A mobile scaffold is put together by slotting sections until the required height is reached. The working platform needs to have a suitable edge protection such as guard-rails and toe-boards. Always push from the bottom of the base and not from the top to move it, otherwise it may lean or topple over.

Figure 1.8 A tower scaffold

WORKING SAFELY WITH ELECTRICITY

It is essential whenever you work with electricity that you are competent and that you understand the common dangers. Electrical tools must be used in a safe manner on site. There are precautions that you can take to prevent possible injury, or even death.

Precautions

Whether you are using electrical tools or equipment on site, you should always remember the following:

* Use the right tool for the job.

* Use a transformer with equipment that runs on 110V.

* Keep the two voltages separate from each other. You should avoid using 230V where possible but, if you must, use a residual current device (RCD) if you have to use 230V.

* When using 110V, ensure that leads are yellow in colour.

* Check the plug is in good order

* Confirm that the fuse is the correct rating for the equipment.

* Check the cable (including making sure that it does not present a tripping hazard).

* Find out where the mains switch is, in case you need to turn off the power in the event of an emergency.

* Never attempt to repair electrical equipment yourself.

* Disconnect from the mains power before making adjustments, such as changing a drill bit.

* Make sure that the electrical equipment has a sticker that displays a recent test date.

Visual inspection and testing is a three-stage process:

1. The user should check for potential danger signs, such as a frayed cable or cracked plug.

2. A formal visual inspection should then take place. If this is done correctly then most faults can be detected.

3. Combined inspections and **PAT** should take place at regular intervals by a competent person.

Watch out for the following causes of accidents – they would also fail a safety check:

KEY TERMS

PAT

– Portable Appliance Testing – regular testing is a health and safety requirement under the Electricity at Work Regulations (1989).

* damage to the power cable or plug

* taped joints on the cable

* wet or rusty tools and equipment

* weak external casing

* loose parts or screws

* signs of overheating

* the incorrect fuse

* lack of cord grip

* electrical wires attached to incorrect terminals

* bare wires.

When preparing to work on an electrical circuit, do not start until a permit to work has been issued by a supervisor or manager to a competent person.

Make sure the circuit is broken before you begin. A 'dead' circuit will not cause you, or anybody else, harm. These steps must be followed:

* Switch off – ensure the supply to the circuit is switched off by disconnecting the supply cables or using an isolating switch.

* Isolate – disconnect the power cables or use an isolating switch.

* Warn others – to avoid someone reconnecting the circuit, place warning signs at the isolation point.

* Lock off – this step physically prevents others from reconnecting the circuit.

* Testing – is carried out by electricians but you should be aware that it involves three parts:

   1. testing a voltmeter on a known good source (a live circuit) so you know it is working properly

   2. checking that the circuit to be worked on is dead

   3. rechecking your voltmeter on the known live source, to prove that it is still working properly.

It is important to make sure that the correct point of isolation is identified. Isolation can be next to a local isolation device, such as a plug or socket, or a circuit breaker or fuse.

The isolation should be locked off using a unique key or combination. This will prevent access to a main isolator until the work has been completed. Alternatively, the handle can be made detachable in the OFF position so that it can be physically removed once the circuit is switched off.

Dangers

You are likely to encounter a number of potential dangers when working with electricity on construction sites or in private houses. Table 1.11 outlines the most common dangers.

Danger	Identifying the danger
Faulty electrical equipment	Visually inspect for signs of damage. Equipment should be double insulated or incorporate an earth cable.
Damaged or worn cables	Check for signs of wear or damage regularly. This includes checking power tools and any wiring in the property.
Trailing cables	Cables lying on the ground, or worse, stretched too far, can present a tripping hazard. They could also be cut or damaged easily.
Cables and pipe work	Always treat services you find as though they are live. This is very important as services can be mistaken for one another. You may have been trained to use a cable and pipe locator that finds cables and metal pipes.
Buried or hidden cables	Make sure you have plans. Alternatively, use a cable and pipe locator, mark the positions, look out for signs of service connection cables or pipes and hand-dig trial holes to confirm positions.
Inadequate over-current protection	Check circuit breakers and fuses are the correct size current rating for the circuit. A qualified electrician may have to identify and label these.

Table 1.11 Common dangers when working with electricity

Each year there are around 1,000 accidents at work involving electric shocks or burns from electricity. If you are working in a construction site you are part of a group that is most at risk. Electrical accidents happen when you are working close to equipment that you think is disconnected but which is, in fact, live.

Another major danger is when electrical equipment is either misused or is faulty. Electricity can cause fires and contact with the live parts can give you an electric shock or burn you.

Different voltages

The two most common voltages that are used in the UK are 230V and 110V:

* 230V: this is the standard domestic voltage. But on construction sites it is considered to be unsafe and therefore 110V is commonly used.

* 110V: these plugs are marked with a yellow casement and they have a different shaped plug. A transformer is required to convert 230V to 110V.

Some larger homes, as well as industrial and commercial buildings, may have 415V supplies. This is the same voltage that is found on overhead electricity cables. In most houses and other buildings the voltage from these cables is reduced to 230V. This is what most electrical equipment works from. Some larger machinery actually needs 415V.

In these buildings the 415V comes into the building and then can either be used directly or it is reduced so that normal 230V appliances can be used.

Colour coded cables

Normally you will come across three differently coloured wires: Live, Neutral and Earth. These have standard colours that comply with European safety standards and to ensure that they are easily identifiable. However, in some older buildings the colours are different.

Wire type	Modern colour	Older colour
Live	Brown	Red
Neutral	Blue	Black
Earth	Yellow and Green	Yellow and Green

Table 1.12 Colour coding of cables

Working with equipment with different electrical voltages

You should always check that the electrical equipment that you are going to use is suitable for the available electrical supply. The equipment's power requirements are shown on its rating plate. The voltage from the supply needs to match the voltage that is required by the equipment.

Storing electrical equipment

Electrical equipment should be stored in dry and secure conditions. Electrical equipment should never get wet but – if it does happen – it should be dried before storage. You should always clean and adjust the equipment before connecting it to the electricity supply.

PERSONAL PROTECTIVE EQUIPMENT (PPE)

Personal protective equipment, or PPE, is a general term that is used to describe a variety of different types of clothing and equipment that aim to help protect against injuries or accidents. Some PPE you will use on a daily basis and others you may use from time to time. The type of PPE you wear depends on what you are doing and where you are. For example, the practical exercises in this book were photographed at a college, which has rules and requirements for PPE that are different to those on large construction sites. Follow your tutor's or employer's instructions at all times.

Types of PPE

PPE literally covers from head to foot. Here are the main PPE types.

Figure 1.9 A hi-vis jacket

Figure 1.10 Safety glasses and goggles

Figure 1.11 Hand protection

Figure 1.12 Head protection

Figure 1.13 Hearing protection

Protective clothing

Clothing protection such as overalls:

* provides some protection from spills, dust and irritants
* can help protect you from minor cuts and abrasions
* reduces wear to work clothing underneath.

Sometimes you may need waterproof or chemical-resistant overalls.

High visibility (hi-vis) clothing stands out against any background or in any weather conditions. It is important to wear high visibility clothing on a construction site to ensure that people can see you easily. In addition, workers should always try to wear light-coloured clothing underneath, as it is easier to see.

You need to keep your high visibility and protective clothing clean and in good condition.

Employers need to make sure that employees understand the reasons for wearing high visibility clothing and the consequences of not doing so.

Eye protection

For many jobs, it is essential to wear goggles or safety glasses to prevent small objects, such as dust, wood or metal, from getting into the eyes. As goggles tend to steam up, particularly if they are being worn with a mask, safety glasses can often be a good alternative.

Hand protection

Wearing gloves will help to prevent damage or injury to the hands or fingers. For example, general purpose gloves can prevent cuts, and rubber gloves can prevent skin irritation and inflammation, such as contact dermatitis caused by handling hazardous substances. There are many different types of gloves available, including specialist gloves for working with chemicals.

Head protection

Hard hats or safety helmets are compulsory on building sites. They can protect you from falling objects or banging your head. They need to fit well and they should be regularly inspected and checked for cracks. Worn straps mean that the helmet should be replaced, as a blow to the head can be fatal. Hard hats bear a date of manufacture and should be replaced after about 3 years.

Hearing protection

Ear defenders, such as ear protectors or plugs, aim to prevent damage to your hearing or hearing loss when you are working with loud tools or are involved in a very noisy job.

Respiratory protection

Breathing in fibre, dust or some gases could damage the lungs. Dust is a very common danger, so a dust mask, face mask or respirator may be necessary.

Make sure you have the right mask for the job. It needs to fit properly otherwise it will not give you sufficient protection.

Foot protection

Foot protection is compulsory on site, particularly if you are undertaking heavy work. Footwear should include steel toecaps (or equivalent) to protect feet against dropped objects, midsole protection (usually a steel plate) to protect against puncture or penetration from things like nails on the floor and soles with good grip to help prevent slips on wet surfaces.

Figure 1.14 Respiratory protection

Legislation covering PPE

The most important piece of legislation is the Personal Protective Equipment at Work Regulations (1992). It covers all sorts of PPE and sets out your responsibilities and those of the employer. Linked to this are the Control of Substances Hazardous to Health (2002) and the Provision and Use of Work Equipment Regulations (1992 and 1998).

Storing and maintaining PPE

All forms of PPE will be less effective if they are not properly maintained. This may mean examining the PPE and either replacing or cleaning it, or if relevant testing or repairing it. PPE needs to be stored properly so that it is not damaged, contaminated or lost. Each type of PPE should have a CE mark. This shows that it has met the necessary safety requirements.

Importance of PPE

PPE needs to be suitable for its intended use and it needs to be used in the correct way. As a worker or an employee you need to:

* make sure you are trained to use PPE

* follow your employer's instructions when using the PPE and always wear it when you are told to do so

* look after the PPE and if there is a problem with it report it.

Your employer will:

* know the risks that the PPE will either reduce or avoid

* know how the PPE should be maintained

* know its limitations.

Consequences of not using PPE

The consequences of not using PPE can be immediate or long-term. Immediate problems are more obvious, as you may injure yourself. The longer-term consequences could be ill health in the future. If your employer has provided PPE, you have a legal responsibility to wear it.

FIRE AND EMERGENCY PROCEDURES

KEY TERMS

Assembly point

– an agreed place outside the building to go to if there is an emergency.

If there is a fire or an emergency, it is vital that you raise the alarm quickly. You should leave the building or site and then head for the **assembly point.**

When there is an emergency a general alarm should sound. If you are working on a larger and more complex construction site, evacuation may begin by evacuating the area closest to the emergency. Areas will then be evacuated one-by-one to avoid congestion of the escape routes.

Three elements essential to creating a fire

Three ingredients are needed to make something combust (burn):

* oxygen   * heat   * fuel.

The fuel can be anything which burns, such as wood, paper or flammable liquids or gases, and oxygen is in the air around us, so all that is needed is sufficient heat to start a fire.

The fire triangle represents these three elements visually. By removing one of the three elements the fire can be prevented or extinguished.

Figure 1.15 Assembly point sign

How fire is spread

Fire can easily move from one area to another by finding more fuel. You need to consider this when you are storing or using materials on site, and be aware that untidiness can be a fire risk. For example, if there are wood shavings on the ground the fire can move across them, burning up the shavings.

Figure 1.16 The fire triangle

Heat can also transfer from one source of fuel to another. If a piece of wood is on fire and is against or close to another piece of wood, that too will catch fire and the fire will have spread.

On site, fires are classified according to the type of material that is on fire. This will determine the type of fire-fighting equipment you will need to use. The five different types of fire are shown in Table 1.13.

Class of fire	Fuel or material on fire
A	Wood, paper and textiles
B	Petrol, oil and other flammable liquids
C	LPG, propane and other flammable gases
D	Metals and metal powder
E	Electrical equipment

Table 1.13 Different classes of fire

There is also F, cooking oil, but this is less likely to be found on site, except in a kitchen.

Taking action if you discover a fire and fire evacuation procedures

During induction, you will have been shown what to do in the event of a fire and told about assembly points. These are marked by signs and somewhere on the site there will be a map showing their location.

If you discover a fire you should:

* sound the alarm

* not attempt to fight the fire unless you have had fire marshal training

* otherwise stop work, do not collect your belongings, do not run, and do not re-enter the site until the all clear has been given.

Different types of fire extinguishers

Extinguishers can be effective when tackling small localised fires. However, you must use the correct type of extinguisher. For example, putting water on an oil fire could make it explode. For this reason, you should not attempt to use a fire extinguisher unless you have had proper training.

When using an extinguisher it is important to remember the following safety points:

* Only use an extinguisher at the early stages of a fire, when it is small.

* The instructions for use appear on the extinguisher.

* If you do choose to fight the fire because it is small enough, and you are sure you know what is burning, position yourself between the fire and the exit, so that if it doesn't work you can still get out.

Type of fire risk	Fire class Symbol	White label Water	Cream label Foam	Black label Carbon dioxide	Blue label Dry powder	Yellow label Wet chemical
A – Solid (e.g. wood or paper)	A	✓	✓	✗	✓	✓
B – Liquid (e.g. petrol)	B	✗	✓	✓	✓	✗
C – Gas (e.g. propane)	C	✗	✗	✓	✓	✗
D – Metal (e.g. aluminium)	D METAL	✗	✗	✗	✓	✗
E – Electrical (i.e. any electrical equipment)	E	✗	✗	✓	✓	✗
F – Cooking oil (e.g. a chip pan)	F	✗	✗	✗	✗	✓

Table 1.14 Types of fire extinguishers

There are some differences you should be aware of when using different types of extinguisher:

* *CO$_2$ extinguishers* – do not touch the nozzle; simply operate by holding the handle. This is because the nozzle gets extremely cold when ejecting the CO$_2$, as does the canister. Fires put out with a CO$_2$ extinguisher may reignite, and you will need to ventilate the room after use.

* *Powder extinguishers* – these can be used on lots of kinds of fire, but can seriously reduce visibility by throwing powder into the air as well as on the fire.

SIGNS AND SAFETY NOTICES

In a well-organised working environment safety signs will warn you of potential dangers and tell you what to do to stay safe. They are used to warn you of hazards. Their purpose is to prevent accidents. Some will tell you what to do (or not to do) in particular parts of the site and some will show you where things are, such as the location of a first aid box or a fire exit.

Types of signs and safety notices

There are five basic types of safety sign, as well as signs that are a combination of two or more of these types. These are shown in Table 1.15.

Type of safety sign	What it tells you	What it looks like	Example
Prohibition sign	Tells you what you must *not* do	Usually round, in red and white	Do not use ladder
Hazard sign	Warns you about hazards	Triangular, in yellow and black	Caution Slippery floor
Mandatory sign	Tells you what you *must* do	Round, usually blue and white	Masks must be worn in this area
Safe condition or information sign	Gives important information, e.g. about where to find fire exits, assembly points or first aid kit, or about safe working practices	Green and white	First aid
Firefighting sign	Gives information about extinguishers, hydrants, hoses and fire alarm call points, etc.	Red with white lettering	Fire alarm call point
Combination sign	These have two or more of the elements of the other types of sign, e.g. hazard, prohibition and mandatory		DANGER Isolate before removing cover

Table 1.15 Different types of safety signs

TEST YOURSELF

1. Which of the following requires you to tell the HSE about any injuries or diseases?

   a. HASAWA

   b. COSHH

   c. RIDDOR

   d. PUWER

2. What is a prohibition notice?

   a. An instruction from the HSE to stop all work until a problem is dealt with

   b. A manufacturer's announcement to stop all work using faulty equipment

   c. A site contractor's decision not to use particular materials

   d. A local authority banning the use of a particular type of brick

3. Which of the following is considered a major injury?

   a. Bruising on the knee

   b. Cut

   c. Concussion

   d. Exposure to fumes

4. If there is an accident on a site who is likely to be the first to respond?

   a. First aider

   b. Police

   c. Paramedics

   d. HSE

5. Which of the following is a summary of risk assessments and is used for high risk activities?

   a. Site notice board

   b. Hazard book

   c. Monitoring statement

   d. Method statement

6. Some substances are combustible. Which of the following are examples of combustible materials?

   a. Adhesives

   b. Paints

   c. Cleaning agents

   d. All of these

7. What is dermatitis?

   a. Inflammation of the skin

   b. Inflammation of the ear

   c. Inflammation of the eye

   d. Inflammation of the nose

8. Screens, netting and guards on a site are all examples of which of the following?

   a. PPE

   b. Signs

   c. Perimeter safety

   d. Electrical equipment

9. Which of the following are also known as scissor lifts or cherry pickers?

   a. Bench saws

   b. Hand-held power tools

   c. Cement additives

   d. Mobile elevated work platforms

10. In older properties the neutral electricity wire is which colour?

    a. Black

    b. Red

    c. Blue

    d. Brown

Chapter 2

Unit CSA–L2Core04

UNDERSTAND INFORMATION, QUANTITIES AND COMMUNICATION WITH OTHERS

LEARNING OUTCOMES

LO1: Know how to interpret and produce information relating to construction

LO2: Understand how to estimate quantities of resources

LO3: Understand how to communicate workplace requirements efficiently

INTRODUCTION

The aim of this chapter is to:

* help you interpret and produce information relating to construction

* show you how to estimate quantities of resources

* enable you to communicate workplace requirements effectively to all levels of the construction team.

INTERPRETING AND PRODUCING INFORMATION

Even quite simple construction projects will require documents. These provide you with the necessary information that you will need to do the job. The documents are produced by a range of different people and each document has a different purpose. Together they give you the full picture of the job, from the basic outline through to the technical specifications.

Types of supporting information

Supporting information can be found in a variety of different types of documents. These include:

* drawings and plans

* programmes of work

* procedures

* specifications

* policies

* schedules

* manufacturers' technical information

* organisational documentation

* training and development records

* risk and method statements

* Construction (Design and Management) (CDM) Regulations

* Building Regulations.

Drawings and plans

Drawings are an important part of construction work. You will need to understand how drawings provide you with the information required to carry out the work. The drawings show what the building will look like and how it will be constructed. This means that there are several different drawings of the building from different viewpoints. In practice, most of the drawings are shown on the same sheet.

Block plans

Block plans show the construction site and the surrounding area. Normally block plans are at a ratio of 1:2500 and 1:1250. This means that 1 millimetre on a block plan is equal to 2,500 mm (2.5 m) or 1,250 mm (1.25 m) on the ground.

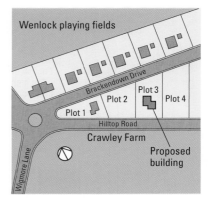

Figure 2.1 Block plan

Site plan

Location drawings are sometimes known as site plans. The site plan drawing shows what is basically planned for the site. It is an important drawing because it has been created in order to get approval for the project from planning committees or funding sources. In most cases the site plan is actually an architectural plan, showing the basic arrangement of buildings and any landscaping.

The site plan will usually show:

* directional orientation (i.e. the north point)

* location and size of the building or buildings

* existing structures

* clear measurements

* colours and materials to be used.

General location

Location drawings show the site or building in relation to its surroundings. It will therefore show details such as boundaries, other buildings and roads. It will also contain other vital information, including:

* access

* drainage

* sewers

* the north point.

The drawing will have a title and will show the scale. A job or project number will help to identify it easily, and it will also have an address, the date when the drawing was done and the name of the client. A version number will also be on the drawing, with an amendment date if there have been any changes. It is important to make sure you have the latest drawing.

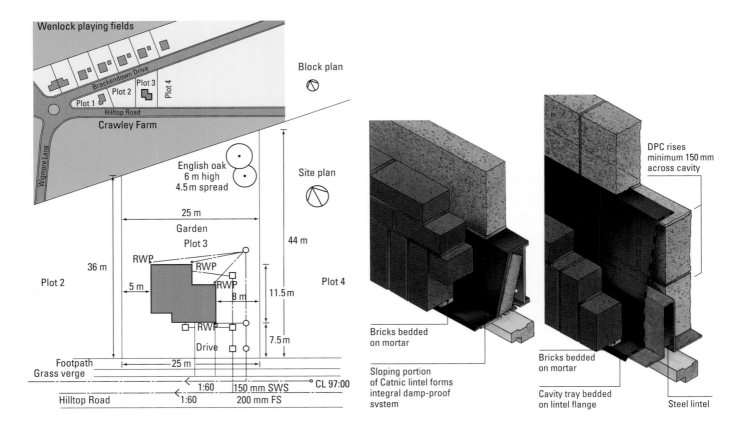

Figure 2.2 Location plan

Figure 2.3 Assembly drawing

Normally location drawings are either 1:500 or 1:200 (that is, 1 mm of the drawing represents 500 mm or 200 mm on the ground).

Assembly

These are detailed drawings that illustrate the different elements and components of the construction. They are likely to be 1:20, 1:10 or 1:5 (1 mm of the drawing represents 20 mm, 10 mm or 5 mm on the ground). This larger scale allows more detail to be shown, to ensure accurate construction.

Sectional

These drawings aim to provide:

* vertical dimensions

* constructional details

* horizontal sections.

They can be used to show the height of ground levels, damp-proof courses, foundations and other aspects of the construction.

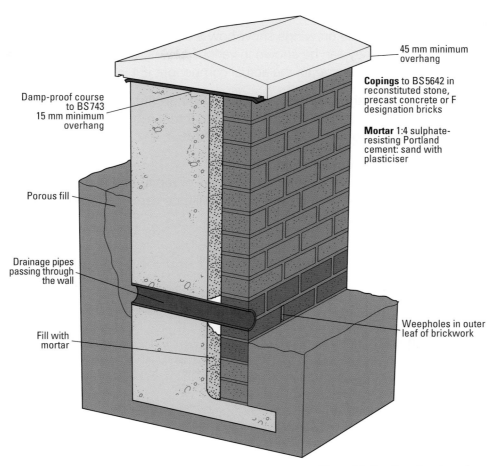

45 mm minimum
overhang

Copings to BS 5642 in
reconstituted stone,
precast concrete or F
designation bricks

Mortar 1:4 sulphate-
resisting Portland
cement: sand with
plasticiser

Damp-proof course
to BS 743
15 mm minimum
overhang

Porous fill

Drainage pipes
passing through
the wall

Fill with
mortar

Weepholes in outer
leaf of brickwork

Figure 2.4 Section drawing of an earth retaining wall

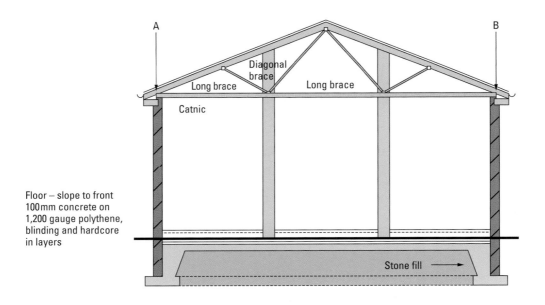

A

B

Diagonal
brace

Long brace

Long brace

Catnic

Floor – slope to front
100 mm concrete on
1,200 gauge polythene,
blinding and hardcore
in layers

Stone fill →

Figure 2.5 Section drawing of a garage

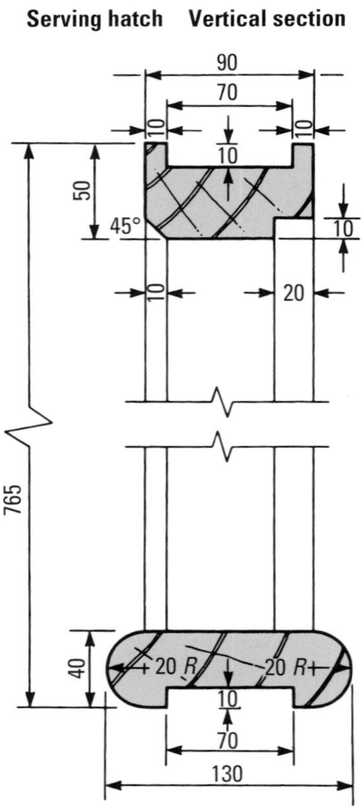

Serving hatch   Vertical section

Figure 2.6 Detail drawing

Details

These drawings show how a component needs to be manufactured. They can be shown in various scales, but mainly 1:10, 1:5 and 1:1 (the same size as the actual component if it is small).

Orthographic projection (first angle)

First angle projection is a view that represents the side of the object as if you were standing away from it, as can be seen in Fig 2.7.

Isometric projection

Isometric projection is a way of representing three-dimensional objects in two dimensions, as can also be seen in Fig 2.7. All horizontal lines are drawn at 30°.

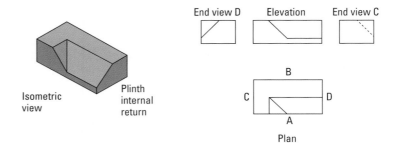

Isometric view

Plinth internal return

End view D   Elevation   End view C

Plan

Figure 2.7 First angle projection

Programmes of work

Programmes of work show the actual sequence of any work activities on a construction project. Part of the work programme plan is to show target times. They are usually shown in the form of a bar or Gantt chart (a special kind of bar chart), as can be seen in Fig 2.8.

In this figure:

* on the left hand side all of the tasks are listed – note this is in logical order

* on the right the blocks show the target start and end date for each of the individual tasks

* the timescale can be days, weeks or months.

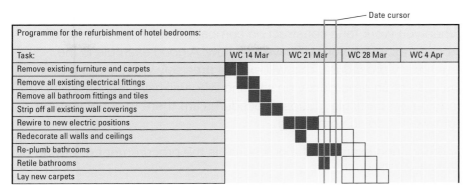

Figure 2.8 Single line contract plan Gantt chart

Far more complex forms of work programmes can also be created. The Gantt chart shown below (Fig 2.9) shows the construction of a house.

This is a more complex example of a bar chart:

* There are two lines – they show the target dates and actual dates. The actual dates are shaded, showing when the work actually began and how long it actually took.

* If this bar chart is kept up to date an accurate picture of progress and estimated completion time can be seen.

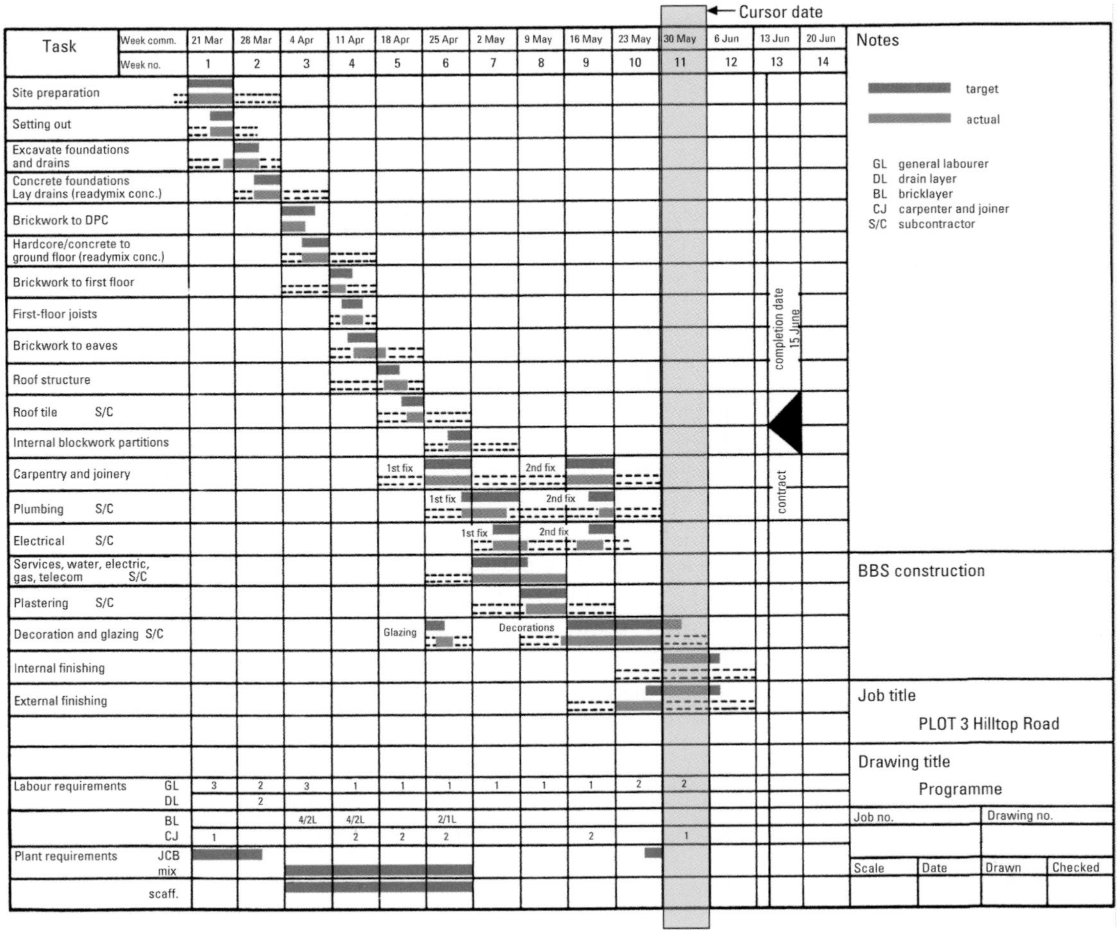

Figure 2.9 Gantt chart for the construction of a house

Procedures

When you work for a construction company it will have a series of procedures which you will have to follow. A good example is the emergency procedure. This will explain precisely what is required in the case of an emergency on site and who will have responsibility for carrying out particular duties. Procedures are there to show you the right way of doing something.

Another good example of a procedure is the procurement or buying procedure. This will outline:

* who is authorised to buy what, and how much individuals are allowed to spend

* any forms or documents that have to be completed when buying.

Specifications

In addition to drawings it is usually necessary to have documents known as specifications. These provide much more information, as can be seen in Fig 2.10.

The specifications give you a precise description. They will include:

* the address and description of the site

* on-site services (e.g. water and electricity)

* materials description, outlining the size, finish, quality and tolerances

* specific requirements, such as the individual who will authorise or approve work carried out

* any restrictions on site, such as working hours.

Policies

Policies are sets of principles or a programme of actions. These are two good examples:

* Environmental policy – how the business goes about protecting the environment.

* Safety policy – how the business deals with health and safety matters and who is responsible for monitoring and maintaining it.

You will normally find both policies and procedures in site rules. These are usually explained to each new employee when they first join the company. Sometimes there may be additional site rules, depending on the job and the location of the work.

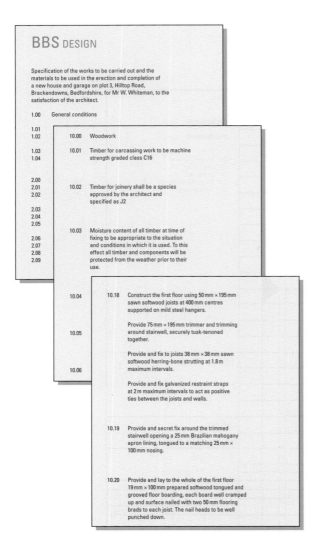

Figure 2.10 Extracts from a typical specification

Schedules

Schedules are cross-referenced to drawings that have been prepared by an architect. They will show specific design information. Usually they are prepared for jobs that will be carried out regularly on site, such as:

* working on windows, doors, floors, walls or ceilings

* working on drainage, lintels or sanitary ware.

A schedule can be seen in Fig 2.11.

The schedule is very useful for a number of purposes, such as:

* working out the quantities of materials needed

* ordering materials and components and then checking them against deliveries

* locating where specific materials will be used.

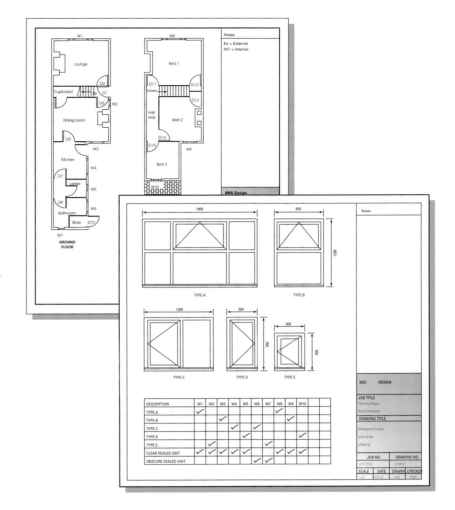

Figure 2.11 Typical windows schedule, range drawing and floor plans

Manufacturers' technical information

Almost everything that is bought to be used on site will come with a variety of types of information. The basic technical information provided will show what the equipment or material is intended to be used for, how it should be stored and any particular requirements it may have, such as for handling or maintenance.

Technical information from the manufacturer can come from a variety of different sources. These may include:

* printed or downloadable data sheets

* printed or downloadable user instructions

* manufacturers' catalogues or brochures

* manufacturers' websites.

Organisational documentation

The potential list of organisational documentation and paperwork is extensive. These are outlined in Table 2.1. Examples can be seen in Figs 2.12 to 2.16.

Document	Purpose
Timesheet	Record of hours that you have worked and the jobs that you have carried out. This is used to help work out your wages and the total cost of the job.
Day worksheet	This details work that has been carried out without providing an estimate beforehand. It usually includes repairs or extra work and alterations.
Variation order	Provided by the architect and given to the builder, showing any alterations, additions or omissions to the original job.
Confirmation notice	Provided by the architect to confirm any verbal instructions.
Daily report or site diary	This covers things that might affect the project like detailed weather conditions, late deliveries or site visitors.
Orders and requisitions	These are order forms, requesting the delivery of materials.
Delivery notes	These are provided by the supplier of materials as a list of all materials being delivered. These need to be checked against materials actually delivered. The buyer will sign the delivery note when they are happy with the delivery.
Delivery records	These are lists of all materials that have been delivered on site.
Memorandum	These are used for internal communications and are usually brief.
Letters	These are used for external communications, usually to customers or suppliers.
Fax	Even though email is commonly used, the industry still uses faxes, as they provide an exact copy of an original document.

Table 2.1 Types of organisational documentation

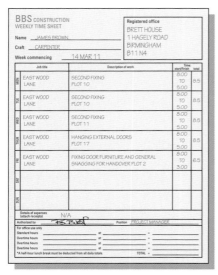

Figure 2.12 Timesheet

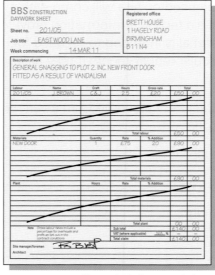

Figure 2.13 Day worksheet

Figure 2.14 Variation order

Training and development records

Training and development is an important part of any job, as it ensures that employees have all the skills and knowledge that they need to do their work. Most medium to large employers will have training policies that set out how they intend to do this.

Employers will have a range of different documents to keep records and to make sure that they are on track. These documents will record all the training that an employee has undertaken.

Training can take place in a number of different ways and different places. It can include:

* induction

* toolbox talks

* in-house training

* specialist training

* training or education leading to formal qualifications.

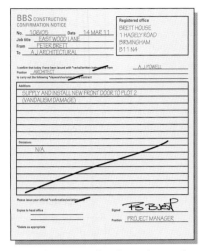

Figure 2.15 Confirmation notice

Checking information for conformity

The information to be checked can include drawings, programmes of work, schedules, policies, procedures, specifications and so on. The term 'conformity' in this sense means:

* making sure that any part of the assembly or component is suitable for the job

* making sure that the standard of work meets the necessary performance requirements.

This may mean that there could be an industry or trade standard that will need to be followed. The actual job or client may also require specific standards.

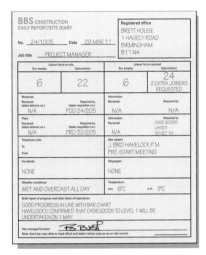

Figure 2.16 Daily report or site diary

Interpreting construction specifications

It would be difficult to put in all of the details in full, so symbols, hatchings and abbreviations are used to simplify the drawings. All of these symbols or hatchings are drawn to follow BS1192. The symbols cover various types of brickwork and blockwork, as well as concrete, hard core and insulation, as can be seen in Fig 2.17.

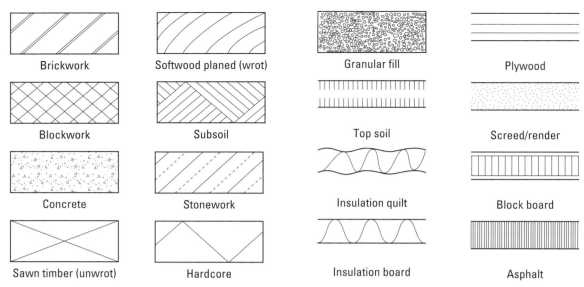

Figure 2.17 Symbols used on drawings

Abbreviation	Meaning
bwk	Refers to all types of brickwork
conc	Refers to areas that will be concreted
dpc	Refers to all types of damp-proof course
fdn	Refers to foundations that are required
insul	Refers to location and type of insulation
rwg	Refers to location of rainwater gulleys
svp	Refers to location and type of soil and vent pipe

Table 2.2 Abbreviations used in construction drawings

Common abbreviations

As we have already seen, covering the drawing with full detail would make it hard to read, so abbreviations are used. Table 2.2 outlines some examples of abbreviations that you will need to become familiar with.

Drawing equipment and its uses

Some basic equipment is necessary in order to produce drawings. These items are outlined in Table 2.3.

Equipment	Explanation and use
Scale rule	This is an essential piece of equipment. It needs to have 1:5/1:50, 1:10/1:100, 1:20/1:200 and 1:250/1:2500.
Set square	You will need to have a pair of these, or an adjustable square. If it is adjustable then you need to be able to create angles of up to 90°. The set square on the shortest side should be at least 150 mm. You will need the ability to create 30, 45, 60 and 90° angles.
Protractor	A protractor is essential to be able to measure angles up to and including 180°.
Compass	Compasses are used to create circles or arcs. It is also advisable to have a divider so that you can easily transfer measurements and dividing lines.
Pencils	For drawings you will need a 2H, 3H or 4H pencil. For sketching and darkening outlines you will need an HB pencil. You will need to keep these sharp.

Table 2.3 Drawing equipment required

In addition to this you will also need at least an A2 size drawing board that has a parallel rule (these may be provided by your college). It is also useful to have an eraser.

Scales used to produce construction drawings

When the plans for individual buildings or construction sites are drawn up they have to be scaled down so that they will fit on a manageable size of paper. It is important to remember that drawings are not sketches and that they are drawn to scale. This means that they are:

* exact and accurate

* in proportion to the real construction.

You can work out the dimensions by using the scale rule when measuring the drawings. There are several common scales used and the measurement is usually metric:

* 1:2500 – the drawing is 2,500 times smaller than the real object

* 1:100 – the drawing is 100 times smaller than the real object

* 1:50 – the drawing is 50 times smaller than the real object

* 1:20 – the drawing is 20 times smaller than the real object

* 1:10 – the drawing is 10 times smaller than the real object

* 1:5 – the drawing is 5 times smaller than the real object

* 1:2 – the drawing is 2 times smaller than the real object.

ESTIMATING QUANTITIES OF RESOURCES

Working out the quantity and cost of resources that are needed to do a particular job is, perhaps, one of the most difficult tasks. In most cases you or the company you work for will be asked to provide a price for the work.

It is generally accepted that there are three ways of doing this:

* estimate – an approximate calculation based on available information

* quotation – which is a fixed price

* tender – tendering is a process of allowing various parties to price for the same work. The process can be open or closed. This usually means that the result is fair.

As we will see a little later in this section, these three ways of costing are very different and each of them has its own problems.

Methods used to estimate quantities

Obviously past experience will help you to quickly estimate the amount of materials that will be needed on particular construction projects. This is also true of working out the best place to buy materials and how much the labour costs will be to get the job finished.

Many businesses will use the *Hutchins UK Building Costs Blackbook*, which provides a construction cost guide. It breaks down all types of work and shows an average cost for each of them.

Computerised estimating packages are available, which will give a comprehensive detailed estimate that looks very professional. This will also help to estimate quantities and timescales.

The alternative is of course to carry out a numerical calculation. It is therefore important to have the right resources upon which to base these calculations. These could be working drawings, schedules or other documents.

Usually all this involves making additions, subtractions, multiplications and divisions. In order to work out the amount of materials you will need for a construction project you will need to know some basic information:

* What does the job entail? How complex is it, and how much labour is required?

* What materials will be used?

* What are the costs of the materials?

Measurement

The standard unit for measurement is the metre (m). There are 100 centimetres (cm) and 1,000 millimetres (mm) in a metre. It is important to remember that drawings and plans have different scales, so these need to be converted to work out the quantities of materials required.

The most basic thing to work out is length (see Fig 2.18), from which you can calculate perimeter, area and then volume, capacity, mass and weight, as can be seen in Table 2.4.

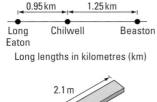

Long lengths in kilometres (km)

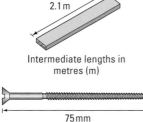

Intermediate lengths in metres (m)

75 mm

Small lengths in millimetres (mm)

Figure 2.18 Length in metres and millimetres

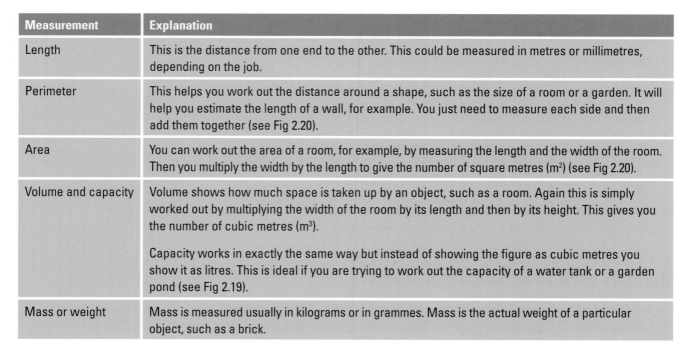

Measurement	Explanation
Length	This is the distance from one end to the other. This could be measured in metres or millimetres, depending on the job.
Perimeter	This helps you work out the distance around a shape, such as the size of a room or a garden. It will help you estimate the length of a wall, for example. You just need to measure each side and then add them together (see Fig 2.20).
Area	You can work out the area of a room, for example, by measuring the length and the width of the room. Then you multiply the width by the length to give the number of square metres (m^2) (see Fig 2.20).
Volume and capacity	Volume shows how much space is taken up by an object, such as a room. Again this is simply worked out by multiplying the width of the room by its length and then by its height. This gives you the number of cubic metres (m^3). Capacity works in exactly the same way but instead of showing the figure as cubic metres you show it as litres. This is ideal if you are trying to work out the capacity of a water tank or a garden pond (see Fig 2.19).
Mass or weight	Mass is measured usually in kilograms or in grammes. Mass is the actual weight of a particular object, such as a brick.

Table 2.4 Working out measurements

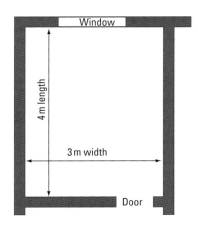

Figure 2.19 Measuring area and perimeter

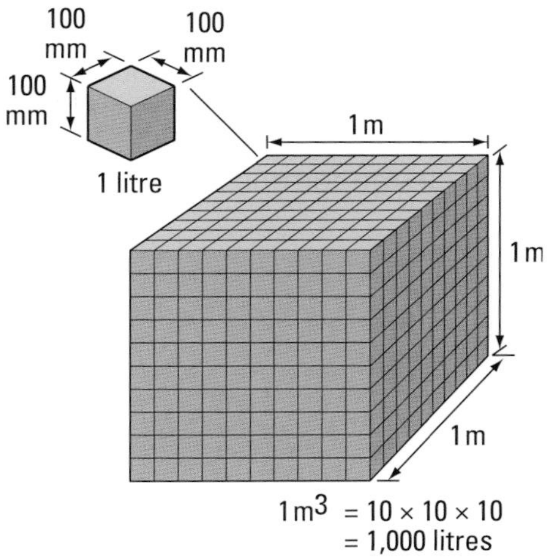

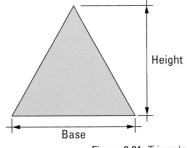

$1 \, m^3 = 10 \times 10 \times 10$
$= 1{,}000 \text{ litres}$

Figure 2.20 Relationship between volume and capacity

Formulae

These can appear to be complicated, but using formulae is essential for working out quantities of materials. Each of the formulae is related to different shapes. In construction work you will often have to work out quantities of materials needed for odd shaped areas.

Area

To work out the area of a triangular shape, you use the following formula:

$$\text{Area (A)} = \text{Base (B)} \times \frac{\text{Height (H)}}{2}$$

So if a triangle has a base of 4.5 and a height of 3.5 the calculation is:

$$4.5 \times \frac{3.5}{2}$$

$$\text{Or } 4.5 \times 3.5 = \frac{15.75}{2} = 7.875 \, m^2$$

Height

If you want to work out the height of a triangle you switch the formulae around. To give:

$$\text{Height} = 2 \times \frac{\text{Area}}{\text{Base}}$$

Perimeter

To work out the perimeter of a rectangle you use the formula:

$$\text{Perimeter} = 2 \times (\text{Length} + \text{Width})$$

It is important to remember this because you need to count the length and the width twice to ensure you have calculated the total distance around the object.

Circles

To work out the circumference or perimeter of a circle you use the formula:

$$\text{Circumference} = \pi \text{ (pi)} \times \text{Diameter}$$

Figure 2.21 Triangle

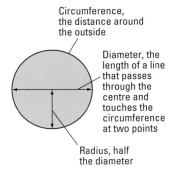

Figure 2.22 Parts of a circle

π (pi) is always the same for all circles and is 3.142.

Diameter is the length of the widest part.

If you know the circumference and need to work out the diameter of the circle the formula is:

$$\text{Diameter} = \frac{\text{Circumference}}{\pi \text{ (pi)}}$$

For example if a circle has a circumference of 15.39 m then to work out the diameter:

$$\frac{15.39}{3.142} = 4.89 \text{ m}$$

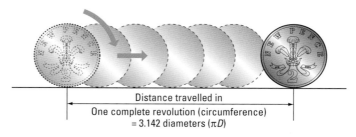

Distance travelled in
One complete revolution (circumference)
= 3.142 diameters (πD)

Figure 2.23 Relationship between circumference and diameter

Complex areas

Land, for example, is rarely square or rectangular. It is made up of odd shapes. You should never feel overwhelmed by complex areas, as all you need to do is to break them down into regular shapes.

By accurately measuring the perimeter you can then break down the shape into a series of triangles or rectangles. All that you need to do then is to work out the area of each of the shapes within the overall shape and add them up together.

Shape		Area equals	Perimeter equals
Square		AA (or A multiplied by A)	4A (or A multiplied by 4)
Rectangle		LB (or L multiplied by B)	2(L+B) (or L plus B multiplied by 2)

Shape		Area equals	Perimeter equals
Trapezium		$\dfrac{(A + B)H}{2}$ (or A plus B multiplied by H and then divided by 2)	A + B + C + D
Triangle		$\dfrac{BH}{2}$ (or B multiplied by H and then divided by 2)	A + B + C
Circle		πr^2 (or r multiplied by itself and then multiplied by pi (3.142))	πd or $2\pi r$

Table 2.5 Calculating complex areas

Volume

Sometimes it is necessary to work out the volume of an object, such as a cylinder or the amount of concrete needed. All that needs to be done is to work out the base area and then multiply that by the height.

For a concrete volume, if a 1.2 m square needs 3 m of height then the calculation is:

$$1.2 \times 1.2 \times 3 = 4.32 \, m^3$$

To work out the volume of a cylinder you need to know the base area × the height. The formula is:

$$\pi r^2 \times H$$

So if a cylinder has a radius (r) of 0.8 and a height of 3.5 m then the calculation is:

$$3.142 \times 0.8 \times 0.8 \times 3.5 = 7.038 \, m^3$$

Pythagoras

Pythagoras' theorem is used to work out the length of the sides of right-angled triangles. It states that:

In all right-angled triangles the square of the longest side is equal to the sum of the squares of the other two sides (that is, the length of a side multiplied by itself).

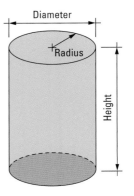

Figure 2.24 Cylinder

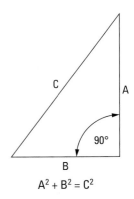

$$A^2 + B^2 = C^2$$

Figure 2.25 Pythagoras' theorem

Measuring materials

Using simple measurements and formulae can help you work out the amount of materials you will need. This is all summarised in Table 2.6.

Material	Measurement
Timber	To work out the linear run of a cubic metre of timber of a given cross sectional area, divide a square metre by the cross sectional area of one piece.
Flooring	To work out the amount of flooring for a particular area in metres² multiply the width of the floor by the length of the floor.
Stud walling, rafters and joists	Measure the distance that the stud partition will cover then divide that distance by a specified spacing and add 1. This will give you the number of spaces between each stud.
Fascias, barges and soffits	Measure the length and then add 10% for waste; however, this will depend on the nearest standard metric size of timber available.
Skirting, dado, picture rails and coving	You need to work out the perimeter of the room and then subtract any doorways or other openings. Again, add 10% for waste.
Bricks and mortar	Half brick walls use 60 bricks per metre squared and one brick walls use double that amount. You should add 5 per cent to take into account any cutting or damage. For mortar assume that you will need 1 kg for each brick.

Table 2.6 Working out materials required

How to cost materials

Once you have found out the quantity of materials necessary, you need to find out the price of those materials. You then do the costing by simply multiplying those prices by the amount of materials actually needed.

CASE STUDY

South Tyneside Homes

South Tyneside Council's Housing Company

It's important to get it right

Glen Campbell is a team leader at South Tyneside Homes.

'Your English and maths skills really are important. As an apprentice, you have to be able to communicate properly – to get information and materials back and forth between tradespeople and yourself, to be able to sit and put a little drawing down, to label things up, and to take information off drawings – especially on the capital works jobs. You're reading and writing stuff down all the time... even your timesheets because they have to be accurate.

When it comes to your maths skills, you're using measurement all the time. If you get measurements wrong, you're not making the money. For example, if you're using the wrong size timber for a roof – the drawing says you've got to use 200 × 50 mm joists and then you go and use ones that are 150 mm – it's either going to cost you more to go back and get it right, or it's not going to be able to take that stress load once the roof goes on. In the end it could even collapse.'

Materials and purchasing systems

Many builders and companies will have preferred suppliers of materials. Many of them will already have negotiated discounts based on their likely spending with that supplier over the course of a year. The supplier will then be organised to supply them at an agreed price.

In other cases, builders may shop around to find the best price for the materials that match the specification. The lowest price may not necessarily be the best one to go for. All materials need to be of a sufficient quality. The other key consideration is whether the materials are immediately available for delivery.

It is vital that suppliers are reliable and that they have sufficient materials in stock. Delays in deliveries can cause major setbacks on site. It is not always possible to warn suppliers that materials will be needed, but a well-run site should be able to anticipate the materials that are needed and put in the orders within good time.

Large quantities may be delivered direct from the manufacturer straight to site. This is preferable when dealing with items where colours must be consistent.

Comparing estimated labour rates

The cost of labour for particular jobs is based on the hourly charge-out rate for that individual or group of individuals multiplied by the time it would take to complete the job.

Labour rates can depend on:

* the expertise of the construction worker

* the size of the business they work for

* the part of the country in which the work is being carried out

* the complexity of the work.

According to the International Construction Costs Survey 2012, the following were average costs per hour:

* Group 1 tradespeople – plumbers, electricians etc. – £30

* Group 2 tradespeople – carpenters, bricklayers etc. – £30

* Group 3 tradespeople – tillers, carpet layers and plasterers – £30

* General labourers – £18

* Site supervisors – £46.

Quotes, estimated prices and tenders

As we have already seen, estimates, quotes and tenders are very different. It is useful to look at these in slightly more detail, as can be seen in Table 2.7 below.

Type of costing	Explanation
Estimate	This needs to be a realistic proposal of how much a job will cost. An estimate is not binding and the client needs to understand that the final cost might be more.
Quote	This is a fixed price based on a fixed specification. The final price may be different if the fixed specification changes, for example if the customer asks for additional work then the price will be higher.
Tender	This is a competitive process. The customer advertises the fact that they want a job done and invites tenders. The customer will specify the specifications and schedules and may even provide the drawings. The companies tendering then prepare their own documents and submit their price based on the information the customer has given them. All tenders are submitted to the customer by a particular date and are either open or closed. The customer then opens all tenders on a given date and awards the contract to the company of their choice. This process is particularly common among public sector customers, such as local authorities.

Table 2.7 Estimates, quotes and tenders

Implications of inaccurate estimates

Larger companies will have an estimating team. Smaller businesses will have someone who has the job of being an estimator. Whenever they are pricing a job, whether it is a quote, an estimate or a tender, they will have to work out the costs of all materials, labour and other costs. They will also have to include a **mark-up**.

It is vital that all estimating is accurate. Everything needs to be measured and checked. All calculations need to be double-checked.

It can be disastrous if these figures are wrong because:

* if the figure is too high then the client is likely to reject the estimate and look elsewhere as some competitors could be cheaper

* if the figure is too low then the job may not provide the business with sufficient profit and it will be a struggle to make any money out of the job.

COMMUNICATING WORKPLACE REQUIREMENTS

Communication can be split into two different types:

* Verbal communication – including face-to-face conversations, discussions in meetings or performance reviews and talking on the telephone.

* Written communication – including all forms of documents, from letters and emails to drawings and work schedules.

KEY TERMS

Mark-up

– a builder or building business, just like any other business, needs to make a profit. Mark-up is the difference between the total cost of the job and the price that the customer is asked to pay for the work.

DID YOU KNOW?

Many businesses fail as a result of not working out their costs properly. They may have plenty of work but they are making very little money.

Each of these forms of communication needs to be clear, accurate and designed in such a way as to make sure that whoever has to use it or refer to it understands it.

Figure 2.26 It's important to communicate effectively, whether it's verbal or written

Key personnel in the communication cycle

Each construction job will require the services of a team of professionals. They have to be able to work and communicate effectively with one another. Each team has different roles and responsibilities. They can be broken down into three particular groups:

* on site   * off site   * visitors.

These are described in Tables 2.8, 2.9 and 2.10.

Role	Responsibilities
Apprentices	They can work for any of the main building services trades under supervision. They only carry out work that has been specifically assigned to them by a trainer, a skilled operative or a supervisor.
Skilled or trade operative	A specialist in a particular trade, such as bricklaying or carpentry. They will be qualified in that trade, or working towards their qualification
Unskilled operatives	Also known as labourers, these are entry level operatives without any formal training. They may be experienced on sites and will take instructions from the supervisor or site manager.
Building services engineers	They are involved in the design, installation and maintenance of heating, water, electrics, lighting, gas and communications. They work either for the main contractor or the architect and give instruction to building services operatives.
Building services operatives	They include all the main trades involved in installation, maintenance and servicing. They take instruction from the building services engineers and work with other individuals, such as the supervisor and charge-hand.
Charge-hand	This person supervises a specific trade, such as carpenters and bricklayers.
Trade foreperson	This person supervises the day-to-day running of the site, and organises the charge-hand and any other operatives.
Site manager	This person runs the construction site, makes plans to avoid problems and meet deadlines, and ensures all processes are carried out safely. They communicate directly with the client.
Supervisor	The supervisor works directly for the site manager on larger projects and carries out some of the site manager's duties on their behalf.
Health and safety officer	This person is responsible for managing the safety and welfare of the construction site. They will carry out inspections, provide training and correct hazards.

Table 2.8 On-site construction team

Role	Responsibilities
Client	The client, such as a local authority, commissions the job. They define the scope of the work and agree on the timescale and schedule of payments.
Customer	For domestic dwellings, the customer may be the same as the client, but for larger projects a customer may be the end user of the building, such as a tenant renting local authority housing or a business renting an office. These individuals are most affected by any work on site. They should be considered and informed with a view to them suffering as little disruption as possible.
Architect	They are involved in designing new buildings, extensions and alterations. They work closely with clients and customers to ensure the designs match their needs. They also work closely with other construction professionals, such as surveyors and engineers.
Consultant	Consultants such as civil engineers work with clients to plan, manage, design or supervise construction projects. There are many different types of consultant, all with particular specialisms.
Main contractor	This is the main business or organisation employed to head up the construction work. The contractor organises the on-site building team and pulls together all necessary expertise. They manage the whole project, taking full responsibility for its progress and costs.
Clerk of works	This person is employed by the architect on behalf of a client. They oversee the construction work and ensure that it represents the interests of the client and follows agreed specifications and designs.
Quantity surveyor	Quantity surveyors are concerned with building costs. They balance maintaining standards and quality against minimising the costs of any project. They need to make choices in line with Building Regulations. They may work either for the client or for the contractor.
Estimator	Estimators calculate detailed cost breakdowns of work based on specifications provided by the architect and main contractor. They work out the quantity and costs of all building materials, plant required and labour costs.
Sub-contractor	They carry out work on behalf of the main contractor and are usually specialist tradespeople or professionals, such as electricians. Essentially, they provide a service and are contracted to complete their part of the project.
Supplier/wholesaler contracts manager	They work for materials suppliers or stockists, providing materials that match required specifications. They agree prices and delivery dates.

Table 2.9 Off-site construction team

Site visitor	Role and responsibility
Training officers and assessors	These people work for approved training providers. They visit the site to observe and talk to apprentices and their mentors or supervisors. They assess apprentices' competence and help them to put together the paperwork needed to show evidence of their skills.
Building control inspector	This person works for the local authority to ensure that the construction work conforms to regulations, particularly the Building Regulations. They check plans, carry out inspections, issue completion certificates, work with architects and engineers and provide technical knowledge on site.
Water inspector	This person carries out checks of plumbing and drainage systems on construction sites.
Health and Safety Executive (HSE) inspector	An HSE inspector can enter any workplace without giving notice. They will look at the workplace, the activities and the management of health and safety to ensure that the site complies with health and safety laws. They can take action if they find there is a risk to health and safety on site.
Electrical services inspector	Inspectors are approved by the National Inspection Council for Electrical Installation Contracting. They check all electrical installation has been carried out in accordance with legislation, particularly Part P of the Building Regulations.

Table 2.10 Construction visitors

Effects of poor communication

Effective communication is essential in all types of work. It needs to be clear and to the point, as well as accurate. Above all it needs to be a two-way process. This means that any communication that you have with anyone must be understood by them. It means thinking before communicating. Never assume that someone understands you unless they have confirmed that they do.

In construction work you have to keep to schedule and work on time, and it is important to follow precise instructions and specifications. Failing to communicate will always cause confusion, extra cost and delays; it can lead to problems with health and safety and accidents. Such problems are unacceptable and very easy to avoid. Negative communication or poor communication can damage the confidence that others have in you to do your job.

Good communication means efficiency and achievement.

> **REED TIP**
>
> As well as within your own team, it is important to communicate clearly with the other trades working on a site, especially if there's a problem that may delay the next stage of the job.

Communication techniques and teamwork

It is important to have a good working relationship with colleagues at work. An important part of this is to communicate in a clear way with them. This helps everyone understand what is going on and what decisions have been made. It also means being clear. Most communication with colleagues will be verbal (spoken). Good communication means:

* cutting out mistakes and stoppages (saving money)

* avoiding delays

* making sure that the job is done right the first time and every time.

Figure 2.27 A water inspection

Equality and diversity in communication

Equality and diversity is not simply about treating everyone in the same way. It is actually recognising that people are different and have different needs. Each of us is unique. This could mean that you are working with people of a different culture, a different age (younger or older), or who follow different religions. It might refer to marital status or gender, sexual orientation or your first language.

In all your actions and your communications you should:

* recognise and respect other people's backgrounds

* recognise that everyone has rights and responsibilities

* not harass or be offensive and use language or behaviour that discriminates.

You should also remember that not everyone's first language will be English so they may not understand everything or be able to communicate clearly with you. You might also find that some colleagues may have hearing impairments (or may not hear what you're saying because they are in a noisy environment). In cases like these, use simple language and check that both you and the person you are communicating with have understood the message.

CASE STUDY

South Tyneside Homes

South Tyneside Council's Housing Company

Using writing and maths in the real world

Gary Kirsop, Head of Property Services, says:

'People seem to think that trades are all about your hands, but it's more than that. You're measuring complicated things – all the trades need to have about the same technical level for planning, calculation and writing reports. You need that level to get through your exams for the future too. When you have one day a week in college, but four days a week working with customers in the real world, without communications skills, it would all fall apart. You have to understand that people come from different backgrounds and that they have their own communication modes. Having good GCSEs will really help you get by in the trade.'

Advantages and disadvantages of different methods of communication

As you progress in your career in construction, you may come across a number of different documents that are used either in the workplace or are provided to customers or clients. All of these documents have a specific purpose. Their exact design may vary from business to business, but the information contained on them will usually be similar.

Documents in the workplace

This group of documents tend to be used only within the workplace. Their general purpose is to collect information or to pass on information from one part of the business to another.

Document type	Purpose
Job specifications	These are detailed sets of requirements that cover the construction, features, materials, finishes and performance specifications required for each major aspect of a project. They may, for example, require a particular level of energy efficiency.
Plans or drawings	These are prepared by architects. They are drawn to scale and provide a standard detailed drawing. They will be used as blueprints (instructions) by building services engineers and operatives while they are working on the site.
Work programmes	These are detailed breakdowns of the order in which work needs to be completed, along with an estimate as to how long each stage is likely to take. For example, a certain amount of time will be allocated for site preparation and then piling and the construction of the substructure of the building. The work programme will indicate when particular skills will be needed and for approximately how long.
Purchase orders	These are documents issued by the buyer to a supplier. They detail the type of materials, quantity and the agreed price so form a record of what has been agreed. The order for materials will have been discussed with the supplier before the purchase order is completed. Many purchase orders are now transmitted electronically, although paper records may be necessary for future reference.
Delivery notes	These are issued to the buyer by the supplier. They act as a checklist for the buyer to ensure that every item requested on the purchase order has been delivered. The buyer will sign the delivery note when they are satisfied with the delivery.
Timesheets	These are completed by those working on site and are verified by the charge-hand, site manager or supervisor. They detail the start and finish times of each individual working on site. They form the basis of the pay calculation for that worker and the overall time that the job has taken.
Policy documents	These cover health and safety, environmental or customer service issues, among others. They outline the requirements of all those working on the site. They will identify roles and responsibilities, codes of conduct or practice, and methods and remedies for dealing with problems or breaches of policy.

Table 2.11 Documents used in the workplace

Documents for customers and clients

Some documents need to be provided to customers and clients. They are necessary to pass on information and can include records of costs and charges that the customer or client is expected to pay for work carried out. Table 2.12 describes what these documents are and their purpose.

Document	Purpose
Quotations and tenders	Quotations provide written details of the costs of carrying out a particular job. They are based on the specification or requirements of the customer or client. They will usually be written by the main contractor on larger sites. A tender is usually a sealed quotation submitted by a contractor at the same time as tenders from other firms in the hope that their quotation will not only match the requirements but will also be the cheapest and therefore the most likely to win the work.
Estimates	An estimate differs from a quotation because it is not a binding quote but a calculation of the cost based on what the contractor thinks the work may involve.
Invoices	An invoice is a list of materials or services that have been provided. Each has an itemised cost and the total is shown at the bottom of the document, along with any additional charges such as VAT.

Document	Purpose
Account statements	This is a record of all the transactions (invoices and payments) made by a customer or client over a given period. It matches payments by the customer and client against invoices raised by the supplier. It also notes any money still owing or over-payments that may have been made.
Contracts	A contract is a legally binding agreement, usually between a contractor and a customer or client, which states the obligations of both parties. A series of agreements are made as part of the contract. It binds both parties to stick to the agreement, which may detail timescales, level of work or costs.
Contract variations	Contract variations are also legally binding. They may be required if both the supplier and the customer or client agrees to change some of the terms of the original contract. This could mean, for example, additional obligations, renegotiating prices or new timescales.
Handover information	Once a project, such as an installation, has been completed, the installer that commissioned the installation will check that it is performing as expected. Handover information includes: • the commissioning document, which details the performance and the checks or inspections that have been made • an installation certificate, which shows that the work has been carried out in accordance with legal requirements and the manufacturer's recommendations.

Table 2.12 Documents used with customers and clients

Other forms of communication

So far we have mainly focused on written forms of communication.

KEY TERMS

VAT

– Value Added Tax is charged on most goods and services. It is charged by businesses or individuals that have raised invoices in excess of £73,000 per year and is currently 20 per cent of the bill.

ITEM	DESCRIPTION	QUANTITY	UNIT	RATE £	AMOUNT £	
	Superstructure: Suspended upper floor					
A	Supply and fit the following C16 grade preservative treated softwood					
A1	50 × 195 mm joists	250	m	6.44	1610	00
A2	75 × 195 mm joists	60	m	8.58	514	80
A3	38 × 150 mm strutting	70	m	3.85	339	50
	Carried to collection:			£	2464	30

Figure 2.28 An example bill of quantities

However one of the most common forms of communication is the telephone, whether landline or mobile. The key advantage of a conversation is that problems and queries can be immediately sorted out. However the biggest problem is that there is no record of any decisions that have been made. It is therefore often wise to ask for written confirmation of anything that has been agreed, perhaps in the form of an email.

Construction is one of the many industries that still prefer to have hard copies of documents. It has been made much easier to send copies of documents as email attachments. The problem though is having an available printer of sufficient quality and size to print off attached documents.

Performance reviews

As you progress through your construction career you will be expected to attend performance reviews. This is another form of communication between you and your immediate supervisor. Certain levels of performance will be expected and will have been agreed at previous reviews. At each review your performance, compared to those standards, will be examined. It gives both sides an opportunity to look at progress. It can help identify areas where you might need additional training or support. It may also show areas of your work that need improvement and more effort from you.

Meetings

Meetings also offer important opportunities for communication. They are usually quite structured and will have a series of topics that form what is known as an **agenda**.

Meetings should give everyone the opportunity to contribute and make suggestions as to how to go forward on particular projects and deal with problems. Individuals are often given the job of preparing information for meetings and then presenting it for discussion. A disadvantage of meetings is that while they are happening construction work is not taking place. This means that it is important to run meetings efficiently and not waste time – but also to ensure that everything that needs to be discussed is covered so that extra meetings do not have to be arranged.

Letters

Today emails have largely overtaken more traditional forms of communication, but letters can still be important. Letters obviously need to be delivered so take longer to arrive than emails but sometimes things do need to be sent through the post. It is polite to put in a **covering letter** with documents or other written communication with clients.

Signs and posters

On a daily basis you will also see a range of signs and posters around larger construction sites. Signs are used to communicate either warnings or information and a full list of different types of sign, particularly those relating to health and safety, can be seen in Chapter 1. Their purpose is to be clear and informative. Posters are often put up in communal areas, such as where you might have lunch or keep your personal belongings. These are designed to be simple and to give you vital information. One disadvantage of signs and posters is that they are a one-way form of communication so if you need more information about them you will need to speak to your supervisor.

REED TIP

A good supervisor will make sure you understand what is expected of you in terms of quality, quantity, the speed of the work and how you'll be working with other trades.

KEY TERMS

Agenda

– a brief list of topics to be discussed at a meeting, outlining any decisions that need to be made.

Covering letter

– this is a very brief letter, often just one paragraph long, which states the purpose of the communication and lists any other documents that have been included.

TEST YOURSELF

1. If a drawing is at a scale of 1:500, each millimetre in the drawing represents how much on the ground?

   a. 1 m

   b. 500 cm

   c. 500 mm

   d. 500 m

2. What is the other term used to describe an orthographic projection?

   a. First angle

   b. Second angle

   c. Assembly drawing

   d. Isometric

3. Which of the following are examples of a manufacturer's technical information?

   a. Data sheets

   b. User instructions

   c. Catalogues

   d. All of these

4. On a drawing, if you were to see the letters FDN, what would that mean?

   a. The signature of the architect

   b. Foundation Design Network

   c. Foundations

   d. Full distance

5. If a drawing is at a scale of 1:5, how many times smaller is the drawing than the real object?

   a. 5 times

   b. 50 times

   c. Half the size

   d. 500 times

6. Which of the following values is pi?

   a. 3.121

   b. 3.424

   c. 3.142

   d. 3.421

7. Which document is used to give detailed sets of requirements that cover the construction, features, materials and finishes?

   a. Work programme

   b. Purchase order

   c. Policy document

   d. Job specification

8. What is VAT?

   a. Volume Added Turnover

   b. Vehicle Attendance Tax

   c. Voluntary Aided Trading

   d. Value Added Tax

9. Which individual on a typical site would sign off timesheets?

   a. Architect

   b. Site manager/supervisor

   c. Delivery driver

   d. Customer

10. Which are the two main types of communication?

    a. Verbal and written

    b. Telephones and emails

    c. Meetings and memorandum

    d. Plans and faxes

Chapter 3

Unit CSA–L2Core05
UNDERSTANDING CONSTRUCTION TECHNOLOGY

LEARNING OUTCOMES

LO1: Understand the principles of foundation construction

LO2: Understand the principles of floor construction

LO3: Understand the principles of wall construction

LO4: Understand the principles of roof construction

LO5: Understand the supply of utilities and services within construction

LO6: Understand the principle of sustainability within construction

INTRODUCTION

The aim of this chapter is to:

* help you understand the range of building materials used within the construction industry

* help you understand their suitability in the construction of modern buildings.

FOUNDATION CONSTRUCTION

Foundations are the primary element of a building as they support and protect the superstructure (the visible part of the building) above. Foundations are part of the substructure of the building, meaning that they are not visible once the building has been completed.

Foundations spread the load of the superstructure and transfer it to the ground below. They provide the building with structural stability and help to protect the building from any ground movement.

Purpose of foundations

It is important to work out the necessary width of foundations. This depends on the total load of the structure and the load-bearing capacity of the ground or subsoil on which the building is being constructed. This means:

* wide foundations are used when the construction is on weak ground, or the superstructure will be heavy

* narrow foundations are used when the subsoil is capable of carrying a heavy weight, or the building is a relatively light load.

The load that is placed on the foundations spreads into the ground at 45°. **Shear failure** will take place if the thickness of the foundations (T) is less than the projection of the wall or column face on the edge of the foundations (P). This is what leads to subsidence (the ground under the structure sinking or collapsing).

As we will see in this section, the depth of the foundation is dependent on the load-bearing capacity of the subsoil. But for the most part foundations should be 200 mm to 300 mm thick.

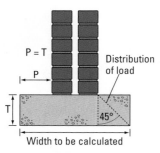

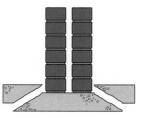

'P' greater than 'T' leads to shear failure

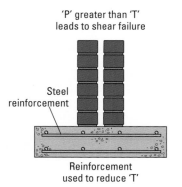

Reinforcement used to reduce 'T'

Figure 3.1 Foundation properties

Different types of foundation

The traditional **strip foundation** is quite narrow and tends to be used for low-rise buildings and dwellings. Most buildings have had unreinforced strip foundations and they were constructed with either brick or block masonry up to the damp course level. Strip foundations can be stepped on sloping ground, in order to cut down on the amount of excavation needed. In poor soil conditions, deep strip foundations can also be used.

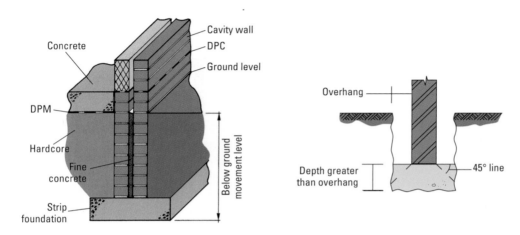

Figure 3.2 Unreinforced strip foundation

Narrow deep strip or trench fill foundations are dug to the foundation depth and then filled with concrete. This reduces excavation, as no bricks or blocks have to be laid into the trench. Trench fill:

* reduces the need to have a wide foundation

* reduces construction time

* speeds up the construction of the footings.

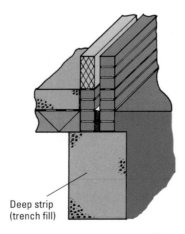

Figure 3.3 Trench fill foundations

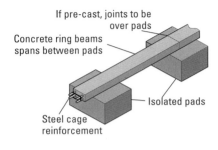

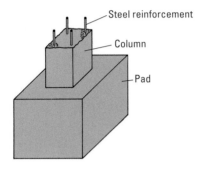

Figure 3.4 Pad foundations

Pad foundations tend to be used for structures that have either a concrete or a steel frame. The pads are placed to support the columns, which transfer the load of the building into the subsoil.

Pile foundations tend to be used for high-rise buildings or where the subsoil is unstable. Holes are bored into the ground and filled with concrete or pre-cast concrete, steel or timber posts are driven into the ground. These piles are then spanned with concrete ring beams with steel reinforcement so that the load of the building is transferred deeper into the ground below. Pile foundations can be short or long depending on how high the building is or how bad the soil conditions are.

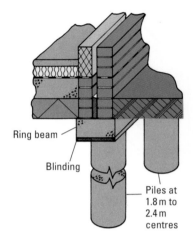

Figure 3.5 Pile foundations

Raft foundations are used when there is a danger that the subsoil is unstable. A large concrete slab reinforced with steel bars is used to outline the whole footprint of the building. It has an edge beam to take the load from the walls, which is transferred over the whole raft. This means that the building effectively 'floats' on the ground surface on top of the concrete raft.

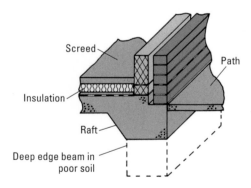

Figure 3.6 Raft foundations

Selecting a foundation

One of the first things that a structural engineer will look at when they investigate a site is the nature of the soil and issues such as the water table (where groundwater begins). The type of soil or ground conditions are very important, as is the possibility of ground movement.

Table 3.1 shows different types of subsoil and how they can affect the choice of foundation.

Subsoil type	Characteristics
Rock	High load bearing but there may be cracks or faults in the rock, which could collapse.
Granular	Medium to high load bearing and can be compacted sand or gravel. If there is a danger of flooding the sand can be washed away.
Cohesive	Low to medium load bearing, such as clay and silt. These are relatively stable, but may have problems with water.
Organic	Low load bearing, such as peat and topsoil. Organic material must be removed before starting the foundations. There is also a great deal of air and water present in the soil.

Table 3.1 Different types of subsoil

The ground may move, particularly if the conditions are wet, extremely dry or there are extremes of temperature. Clay, for example, will shrink in the hot summer months and swell up again in the wet winter months. Frost can affect the water in the ground, causing it to expand.

Ground movement is also affected by the proximity of trees and large shrubs. They will absorb water from the soil, which can dry out the subsoil. This causes the soil underneath the foundations to collapse.

The end use of the building

The other key factor when selecting a foundation is the end use of the building:

* Strip foundation – this is the most common and cheapest type of foundation. Strip foundations are used for low to medium rise domestic and industrial buildings, as the load-bearing will not be high.

* Raft foundation – this is only ever really used when the ground on which the building is being constructed is very soft. It is also sometimes used when the ground across the area is likely to react in different ways because of the weight of the building. In areas of the UK where there has been mining, for example, raft foundations are quite common, as the building could subside. The raft is a rigid, concrete slab reinforced with steel bars. The load of the building is spread across the whole area of the raft.

* Piled foundations – these are used for high-rise buildings where the building will have a high load or where the soil is found to be poor.

Materials used in the construction of foundations

Concrete

Concrete is used to produce a strong and durable foundation. The concrete needs to be poured into the foundation with some care. The size of the foundation will usually determine whether the concrete is actually mixed on site or brought in, in a ready-mixed state, from a supplier. For smaller foundations a concrete mixer and wheelbarrows are usually sufficient. The concrete is then poured into the foundation using a chute.

Concrete consists of both fine and coarse aggregate, along with water, cement and additives if required.

Aggregates

Aggregates are basically fillers. The coarse aggregate is usually either crushed rock or gravel. The grains are 5mm or larger.

Fine aggregate is usually sand that has grains smaller than 5mm.

The fine aggregate fills up any gaps between the particles in the coarse aggregate.

Cement

Cement is an adhesive or binder. It is Portland stone, crushed, burnt and crushed again and mixed with limestone. The materials are powdered and then mixed together to create a fine powder, which is then fired in a kiln.

Water

Potable water, which is water that is suitable for drinking, should be used when making concrete. The reason for this is that drinkable water has not been contaminated and it does not have organic material in it that could rot and cause the concrete to crack. The water mixes with the cement and then coats the aggregate. This effectively bonds everything together.

Additives

Additives, or admixtures, make it possible to control the setting time and other aspects of fresh concrete. It allows you to have greater control over the concrete. Common add mixtures can accelerate the setting time, or reduce the amount of water required. They can:

* give you higher strength concrete

* provide protection against corrosion

* accelerate the time the concrete needs to set

* reduce the speed at which the concrete sets

* provide protection against cracking as the concrete sets (prevent shrinkage)

* improve the flow of the concrete

* improve the finish of the concrete

* provide hot or cold weather protection (a drop or rise in temperature can change the amount of time that a concrete needs to set, so these add mixtures compensate for that).

Reinforcement

Steel bars or mesh can be used to give the foundation additional strength and support. It can also stop the foundation from cracking. Concrete is very good at dealing with loads, so weight coming from above is something concrete can deal with. But when concrete foundations are wide, and parts of them are under tension, there is a danger it may crack.

Concrete should also be levelled, usually with a vibrator or a compactor, although newer types of concrete are self-compacting. All concrete needs to be laid on well-compacted ground.

Figure 3.7 Reinforcement using steel bars (or mesh)

FLOOR CONSTRUCTION

For most domestic buildings floorboards or sheets are laid over timber joists. In other cases, and in most industrial buildings, the ground floors have a block and beam construction with hard core. They then have a damp-proof membrane and over the top is solid concrete.

A floor is a level surface that provides some insulation and carries any loads (for example furniture) and to transfer those loads.

The ground floors have additional purposes. They must stop moisture from entering the building from the ground. They also need to prevent plant or tree roots from entering the building.

Ground floors

For ground floors there are two options:

* Solid – is in contact with the ground.

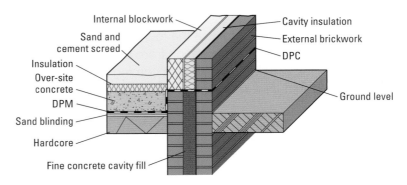

Figure 3.8 Solid ground floors

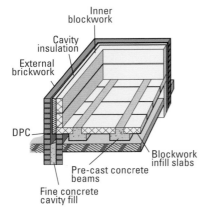

Figure 3.9 Suspended ground floors

* Suspended – the floor does not touch the ground and spans between walls in the building. Effectively there is a void beneath the floor, with air bricks in external walls to allow for ventilation.

The options for ground floors are more complicated than those for upper floors. This is because the ground floors need to perform several functions. It is quite rare for modern buildings to have timber joists and floorboards. Suspended ground floors and traditional timber floors tend to be seen in older buildings. It is far more common to have solid ground floors, or to have timber floors over concrete floors, which are known as floating ground floors.

The key options are outlined in Table 3.2.

Type of floor	Construction and characteristics
Solid	One construction method is to use hard core as the base, with a layer of sand and a layer of insulation such as Celotex, usually 100 mm thick, and then covered with a damp-proof membrane. The concrete is then poured into the foundation. To provide a smooth finish for floor finishes a cement and sand screed is applied, usually after the building has been made watertight..
Timber suspended	A similar process to a solid ground floor is carried out but then, on top of this, dwarf or sleeper walls are built. These are used to support the timber floor. Air bricks are also added to provide necessary ventilation. Joists are then spaced out along the dwarf walls. A damp-proof course is inserted under the floor joists and then floorboards or sheets placed on top of the joists.
Beam and block suspended	Concrete beams and lightweight concrete slabs or blocks are used to create the basic flooring. The beams are evenly spaced across the foundation and gaps between the beams are filled with blocks to form the floor. The blocks and beams are then insulated and it is finished off with either a cement screed or a timber floating floor.
Floating	This timber construction goes over the top of concrete floors. Bearers are put down and then the boarding or sheets are fixed to the bearers. The weight of the boards themselves hold them in place.

Table 3.2 Construction of ground floors

Upper floors

Timber is usually used for these suspended floors in homes and other types of dwellings. In industrial buildings concrete tends to be used.

Timber suspended upper floor

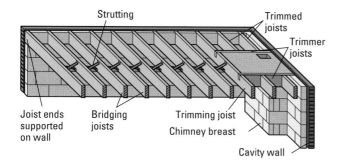

Concrete suspended upper floor

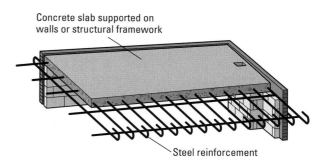

Figure 3.10 Upper floors

For dwellings, bridging joists are used. These are supported at their ends by load-bearing walls. Boarding or sheets provide the flooring for the room on the top of the joists. Underneath the joists plasterboard creates the basis of the ceiling for the room below.

It is also possible to fill the voids between the floorboards and the plasterboard with insulation. Insulation not only helps to prevent heat loss, but can also reduce noise.

Concrete suspended floors are usually either cast on site or available as ready-cast units. They are effectively locked into the structure of the building by steel reinforcement. If the concrete floors are being cast on site then **formwork** is needed. Concrete floors tend to be used in many modern buildings, particularly industrial ones, as they offer greater load bearing capacity, have greater fire resistance and are more sound resistant.

WALL CONSTRUCTION

Walls have a number of different purposes:

* They hold up the roof.

* They provide protection against the elements.

* They keep the occupants of the building warm.

KEY TERMS

Formwork

– this can also be known as shuttering. It is a temporary structure or mould that supports and shapes wet concrete until it cures and is able to be self-supporting.

Many buildings now have double walls, which means as follows:

* The outside wall is a wet one because it is exposed to the elements outside the building.

* The internal wall is dry but it needs to be kept separate from the outside wall by a cavity.

* The cavity or gap acts as a barrier against damp and also provides some heat insulation.

* The cavity can be completely filled or part-filled depending on the insulation value required by Building Regulations.

Within the building there are other walls. These internal walls divide up the space within the building. These do not have to cope with all of the demands of the external walls. As a result, they do not necessarily have to be insulated and are, therefore, thinner. They are block and then covered with plaster. Alternatively they can be a timber framework, which is also known as stud work, and again can be covered with plasterboard.

Different types of wall construction and structural considerations

In addition to walls being external or internal, they can also be classed as being load bearing or non-load bearing.

Internal walls can be either load bearing or non-load bearing. In both internal and external walls, where they are load bearing, any gaps or openings for windows or doors have to be bridged. This is achieved by using either arches or lintels. These support the weight of the wall above the opening.

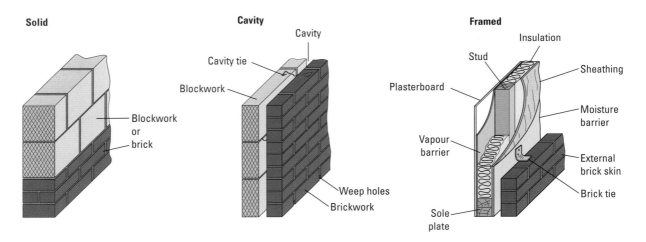

Figure 3.11 Some examples of external wall construction

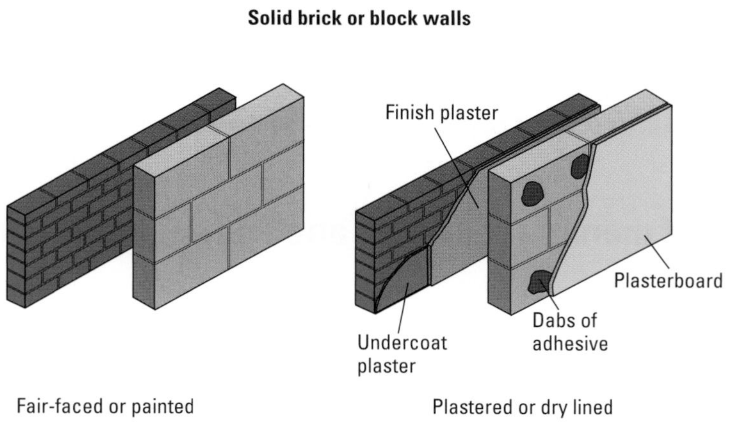

Solid brick or block walls

Timber or metal framed partitions

Finish plaster

Plasterboard

Undercoat plaster

Dabs of adhesive

Fair-faced or painted

Plastered or dry lined

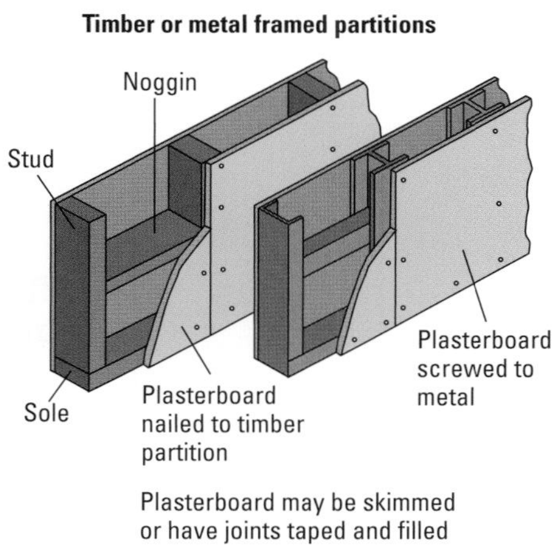

Noggin

Stud

Sole

Plasterboard nailed to timber partition

Plasterboard screwed to metal

Plasterboard may be skimmed or have joints taped and filled

Figure 3.12 Some examples of internal wall construction

Solid masonry

In modern builds solid masonry is quite rare, as it uses up a lot of bricks and blocks. External solid walls tend to be much thinner and made from lightweight blocks in modern builds. They will have some kind of waterproof surface over the top of them, which can be made of render, plastic, metal or timber.

Cavity masonry

As we have seen, cavity walls have an outer and an inner wall and a cavity between them. Usually solid walling, or blockwork, is built up to ground level and then the cavity walling continues to the full height of the building. These cavity walls are ideal for most buildings up to medium height.

Many industrial buildings have cavity walls for the lower part of the building and then have insulated steel panels for the top part of the building.

The usual technique is to have brick for the outer wall and an insulating block for the inner wall. The gap or cavity can then be filled with an insulation material.

Timber framed

Panels made of timber, or in some cases steel, are used to construct walls. They can either be load bearing or non-load bearing and can also be used for the outside of the building or for internal walling. The panels are solid structures and the spaces between the vertical struts (studs) and the horizontal struts (head or sole plates) can be filled with insulation material.

Internal walls or partitions

Internal walls tend to be either solid or framed. Solid walls are made up from blocks. In many industrial buildings the blocks are actually exposed and can be left in their natural state or painted. In domestic buildings plasterboard is usually bonded to the surface and then plastered over to provide a smoother finish.

It is more common for domestic buildings to have framed internal walls, which are known as stud partitions. These are exactly the same as other framed walling, but will usually have plasterboard fixed to them. They would then receive a skimmed coat of plaster to provide the smooth finish.

Damp-proof membrane (DPM) and damp-proof course (DPC)

Damp-proof membranes are installed under the concrete in ground floors in order to ensure that ground moisture does not enter the building. Effectively the membrane waterproofs the building.

Damp-proof courses are a continuation of the damp-proof membrane. They are built into a horizontal course of either block and brickwork, which is a minimum of 150mm above the exterior ground level. DPCs are also designed to stop moisture from coming up from the ground, entering the wall and then getting into the building. The most common DPC is a polythene sheet called visqueen DPC. It comes in rolls to the appropriate width for the wall.

In older buildings lead, bitumen or slate would have been used as a DPC.

ROOF CONSTRUCTION

In a country such as the UK, with a great deal of rain and sometimes snowy weather, it makes sense for roofs to be pitched. Pitched means built at an angle. The idea is that the rain and snow falls down the angle and off the edge of the roof or into gutters rather than lying on the roof.

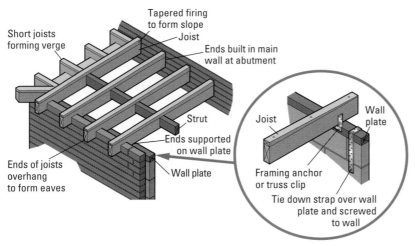

Short joists forming verge

Tapered firing to form slope
Joist

Ends built in main wall at abutment

Strut

Ends supported on wall plate

Ends of joists overhang to form eaves

Wall plate

Joist

Wall plate

Framing anchor or truss clip

Tie down strap over wall plate and screwed to wall

Figure 3.13 Flat roof structure

However, not all roofs are pitched. In fact many domestic dwelling extensions have flat roofs. A great number of industrial buildings have entirely flat roofs. The problem with a flat roof is that it needs to be able to support itself, but just as importantly it needs to be able to carry the additional weight of snow or rain. This means that large flat roofs may have to have steel sections (known as trusses) or even reinforced concrete and beams to increase their load-bearing capacity.

Roofs also provide stability to the walls by tying them together. As we will see, there are several different types of roof. These are usually identified by their pitch or shape.

Types of roof construction

The roof is made up of the rafters and beams. Everything above the framework is regarded as a roof covering, such as slates, tiles and felt.

Table 3.3 outlines some of the key characteristics of different types of roof.

Roof type	Characteristics	How it looks
Flat	This is a roof that has a slope of less than 10°. Generally flat roofs are used for smaller extensions to dwellings and on garages. Traditionally they would have had bitumen felt, although it is becoming more common for fibreglass to be used.	 Figure 3.14
Mono-pitch	This is a roof that has a single sloping surface but is not fixed to another building or wall. The front and back walls could be different heights, or the other exposed surface of the roof is perpendicular.	 Figure 3.15
Gambrel roof	This is a roof that has two differently angled slopes. Usually the upper part of the roof has a fairly shallow pitch or slope and the lower part of the roof has a steeper slope.	 Figure 3.16
Couple roof	This is often called gable end and is one of the most common types of roof for dwellings. A gable is a wall with a triangular upper part. This supports the roof in construction using purlins. This means that the roof has two sloping surfaces, which come down from the ridge to the eaves.	 Figure 3.17
Hipped roof	Hipped roofs have slopes on three or four sides. There are also hipped roofs with single, straight gables.	 Figure 3.18
Lean-to	A lean-to is similar to a mono-pitched roof except it is abutted to a wall. The slope is greater than 10°. The higher part of the roof is fixed to a higher wall.	 Figure 3.19

Table 3.3 Different types of roof

Roofing components

Each visible part of a roof has a specific name and purpose. Table 3.4 explains each of these individual features.

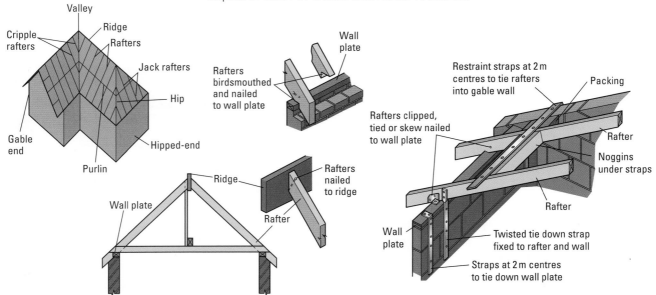

Figure 3.20 Traditional cut roof details

Roof feature	Description
Ridge	This is the top of the roof and the junction of the sloping sides. It is the peak, where the rafters meet.
Purlin	This is a beam that supports the mid-span section of rafters.
Firings	These are angled pieces of timber that are placed on the rafters to create a slope.
Batten	Roof battens are thin strips, usually of wood, which provide a fixing point for either roofing sheets or roof tiles.
Tile	These can be made from clay, slate, concrete or plastic. They are placed in regular, overlapping rows and fixed to the battens.
Fascia	This is a horizontal, decorative board. It is usually a wooden board, although it can be PVC. It is fixed to the ends of the rafters at eaves level and is both a decorative feature and a fixing for rainwater goods.
Wall plate	This is a horizontal timber that is placed at the top of a wall at eaves level. It holds the ends of joists or rafters.
Bracings	Roof rafters need to be braced to make them more rigid and stable. These bracings prevent the roof from buckling. Usually there are several braces in a typical roof.
Felt	Roofing felt has two elements – it has a waterproofing agent (bitumen) and what is known as a carrier. The carrier can be either a polyester sheet or a glass fibre sheet. Roofing felt tends to be used for flat roofs and for roofs with a shallow pitch.
Slate	Slate roofing tiles are usually fixed to timber battens with double nails. They have a lifespan of between 80 and 100 years.
Flashings	Wherever there is a joint or angle on a roof, a thin sheet of either lead or another waterproof material is added. In the past this tended always to be lead. Many different types of flashing can now be used but all have the role of preventing water penetrating into joints, such as on abutments to walls and around the chimney stack.
Rafter	Roof rafters are the main structural components of the roof. They are the framework. They rest on supporting walls. The rafters are set at an angle on sloped roofs or horizontal on a flat roof.
Apex	The apex is the highest point of the roof, usually the ridge line.

Roof feature	Description
Soffit	Soffits are the lower part, or overhanging part, of the eaves. In other words they are the underside of the eaves. A flat section of timber or plastic is usually fixed to the soffit to ensure water tightness.
Bargeboard	This is an ornamental feature, which is fixed to the gable end of a roof in order to hide the ends of roof timbers.
Eaves	These are the area found at the foot of the rafter. They are not always visible as they can be flush. In modern construction, the eaves have two parts: the visible eaves projection and the hidden eaves projection.

Table 3.4 Parts of a roof

Roof coverings

There are many different types of materials that can be used to cover the roof. Even tiles and slates come in a wide variety of shapes and sizes, along with colours and different finishes.

In many cases the type of roof covering is determined by the traditional and local styles in the area. Local authorities want roof coverings that are not too far from the common style in the area. This does not stop manufacturers from coming up with new ideas, however, which can add benefits during construction. There is much innovation and labour saving that also helps to minimise build costs.

Affordable clay tiles, for example, make it possible to use traditional materials that had been out of the budget of many construction jobs for a number of years.

The Table 3.5 outlines some of the more common types of roof covering and describes their main characteristics and use.

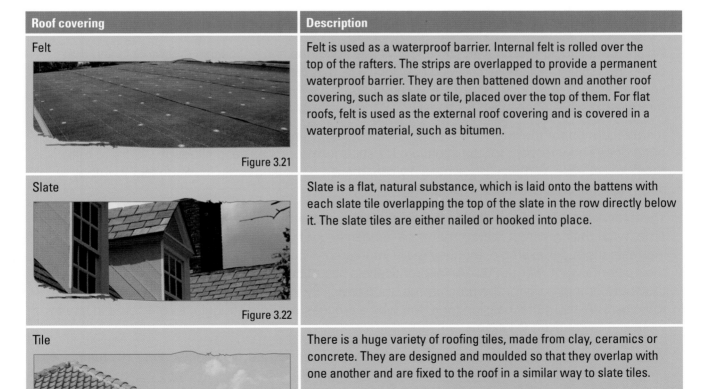

Roof covering	Description
Felt Figure 3.21	Felt is used as a waterproof barrier. Internal felt is rolled over the top of the rafters. The strips are overlapped to provide a permanent waterproof barrier. They are then battened down and another roof covering, such as slate or tile, placed over the top of them. For flat roofs, felt is used as the external roof covering and is covered in a waterproof material, such as bitumen.
Slate Figure 3.22	Slate is a flat, natural substance, which is laid onto the battens with each slate tile overlapping the top of the slate in the row directly below it. The slate tiles are either nailed or hooked into place.
Tile Figure 3.23	There is a huge variety of roofing tiles, made from clay, ceramics or concrete. They are designed and moulded so that they overlap with one another and are fixed to the roof in a similar way to slate tiles.

Roof covering	Description
Metals  Figure 3.24	There are many different types of metal roof covering, such as corrugated sheets, flat sheets, box profile sheets or even sheets that have a tile effect. The metal is galvanised and plastic coated to provide a durable and long-lasting waterproof surface.

Table 3.5 Roof coverings

CASE STUDY

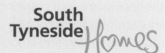

South Tyneside Homes

South Tyneside Council's Housing Company

How to impress in interviews

Andrea Dickson and Gillian Jenkins sit on the interview panels for apprenticeship applications at South Tyneside Homes.

'Interviews are all about the three Ps: Preparation, Presentation and Personality.

An applicant should turn up with some knowledge about the apprenticeship programme and the company itself. For example, knowing how long it is, that they have to go to college and to work – and don't say, "I was hoping you'd tell me about it"! If they've done a bit of research, it will show through and work in their favour – especially if they can explain why it is that they want to work here.

It sets them up for the interview if they come in smartly dressed. We're not marking them on that, but it does show respect for the situation. It's still a formal process and although we try to make them feel at ease as much as we possibly can, there's no getting away from the fact that they're applying for a job and it is a formal setting.

The interviews are a chance to tell the company about themselves: what they do in their spare time, what their greatest achievements have been and why. Applicants should talk about what interests them, for example, are they really interested in becoming a joiner or is that something their parents want them to do? An apprenticeship has to be something they really want to do – if they have enthusiasm for the programme, then they'll fly through it. If not, it's a very long three to four years. Without that passion for it, the whole process will be a struggle; they'll come in late to work and might even fail exams.

We also talk to them about any customer service experiences they've had, working in a team, project working (for example, a time you had to complete a task and what steps you took), as well as asking some questions about health and safety awareness.'

SUPPLY OF UTILITIES AND SERVICES

Most but not all dwellings and other structures are connected in some way to a wide range of utilities and services. In the majority of cities, towns and villages structures are connected to key utilities and services, such as a sewer system, potable (drinking) water, gas and electricity. This is not always the case for more remote structures.

Whenever construction work is carried out, whether it is on an existing structure or a new build, the supply of utilities and services or the linking up of these parts of the **infrastructure** are very important. Often they will require the services of specialist engineers from the **service provider**.

Table 3.6 outlines the main utilities and services that are provided to most structures.

KEY TERMS

Infrastructure

– these are basic facilities, such as a power supply, a road network and a communication link.

Service provider

– these are companies or organisations that provide utilities, such as gas, water, communications or electricity.

Utility or service	Description
Drainage	Drainage is delivered by a range of water and sewerage companies. They are responsible for ensuring that surface water can drain away into their system.
Waste water and sewerage	Any waste water and sewage generated by the occupants of a structure needs to have the necessary pipework to link it to the main sewerage system. It is then sent to a sewage treatment works via the pipework. Remote areas may not be connected to the sewerage system so use septic tanks and cesspools.
Water	Each structure should be linked to the water supply that provides wholesome, potable drinking water. The pipework linking the structure to the water supply needs to be protected to ensure that backflow from any other source does not contaminate the system.
Gas	Each area has a range of different gas suppliers. This is delivered via a service pipe from the main system into the structure. Areas that do not have access to the main gas supply system use gas contained in bottles.
Electricity	The National Grid provides electricity to a variety of different electricity suppliers. It is the National Grid that operates and maintains the cabling. There are around 28 million individual customers in the UK.
Communications (telephone, data, cable)	There are several ways in which telecommunications can be linked to a structure. Traditional telephone poles hold up copper cables and not only provide telephone but also internet access to structures. In cities and many of the larger towns this system is being replaced by cables that are fibre optic and run underground. These are then linked to each individual structure.
Ducting (heating and ventilation)	Heating and ventilation engineers install and maintain duct work. The complex systems are known as HVAC. These systems can transfer air for heating or cooling of the structure. The overall system can also provide hot and cold water systems, along with ventilation.

Table 3.6 Services and utilities

SUSTAINABILITY AND INCORPORATING SUSTAINABILITY INTO CONSTRUCTION PROJECTS

Sustainability is something that we all need to be concerned about as the earth's resources are used up rapidly and climate change becomes an ever-bigger issue. Carbon is present in all fossil fuels, such as coal or natural gas. Burning fossil fuels releases carbon dioxide, which is a greenhouse gas linked to climate change.

Energy conservation aims to reduce the amount of carbon dioxide in the atmosphere. The idea is to do this by making buildings better insulated and, at the same time, make heating appliances more efficient. Sustainability also means attempting to generate energy using renewable and/or low or zero carbon methods.

According to the government's Environment Agency, sustainable construction means using resources in the most efficient way. It also means cutting down on waste on site and reducing the amount of materials that have to be disposed of and put into **landfill**.

In order to achieve sustainable construction the Environment Agency recommends

* reducing construction, demolition and excavation waste that needs to go to landfill

* cutting back on carbon emissions from construction transport and machinery

* responsibly sourcing materials

* cutting back on the amount of water that is wasted

* making sure construction does not have a negative impact on **biodiversity**.

Sustainable construction and incorporating it into construction projects

In the past buildings were generally constructed as quickly as possible and at the lowest cost. More recently the idea of sustainable construction focuses on ensuring that the building is not only of good quality and that it is affordable, but that it is also efficient in terms of energy use and resources.

Sustainable construction also means having the least negative environmental impact. So this means minimising the use of raw materials, energy, land and water. This is not only during the build but also for the lifetime of the building.

Finite and renewable resources

We all know that resources such as coal and oil will eventually run out. These are examples of finite resources.

Oil, however, is not just used as fuel – it is in plastic, dyes, lubricants and textiles. All of these are used in the construction process.

Renewable resources are those that are produced either by moving water, the sun or the wind. Materials that come from plants, such as biodiesel, or the oils used to make adhesives, are all examples of renewable resources.

Figure 3.25 Most modern new-builds follow sustainable principles

The construction process itself is only part of the problem. It is important to consider the longer-term impact and demands that the building will have on the environment. This is why there has been a drive towards sustainable homes and there is a Code for Sustainable Homes.

Construction and the environment

In 2010, construction, demolition and excavation produced 20 million tonnes of waste that had to go into landfill. The construction industry is also responsible for most illegal fly tipping (illegally dumping waste). In any year the Environment Agency responds to around 350 pollution incidents caused as a result of construction.

Regardless of the size of the construction job, everyone in construction is responsible for the impact they have on the environment. Good site layout, planning and management can help to reduce these problems.

Sustainable construction helps to encourage this because it means managing resources in a more efficient way, reducing waste, recycling where possible and reducing your **carbon footprint**.

Architecture and design

The Code for Sustainable Homes Rating Scheme was introduced in 2007. Many local authorities have instructed their planning departments to encourage sustainable development. This begins with the work of the architect who designs the building.

DID YOU KNOW?

Search on the internet for 'sustainable building' and 'improving energy efficiency' to find out more about the latest technologies and products.

KEY TERMS

Carbon footprint

– This is the amount of carbon dioxide produced by a project. This not only includes burning carbon-based fuels such as petrol, gas, oil or coal, but includes the carbon that is generated in the production of materials and equipment.

Local authorities ask that architects and building designers:

* ensure the land is safe for development – so if it is contaminated this is dealt with first

* ensure access to and protect the natural environment – this supports biodiversity and tries to create open spaces for local people

* reduce the negative impact on the local environment – buildings should keep noise, air, light and water pollution down to a minimum

* conserve natural resources and cut back carbon emissions – this covers energy, materials and water

* ensure comfort and security – good access, close to public transport, safe parking and protection against flooding.

Figure 3.26 Sustainable developments aim to be pleasant places to live

Using locally managed resources

The construction industry imports nearly 6 million cubic metres of sawn wood each year. However there is plenty of scope to use the many millions of cubic metres of timber produced in managed forests, particularly in Scotland.

Local timber can be used for a wide variety of different construction projects:

* Softwood – including pines, firs, larch and spruce – for panels, decking, fencing and internal flooring.

* Hardwood – including oak, chestnut, ash, beech and sycamore – for a wide variety of internal joinery.

Eco-friendly, sustainable manufactured products and environmentally resourced timber

There are now many suppliers that offer sustainable building materials as a green alternative. Tiles, for example, are now made from recycled plastic bottles and stone particles.

There is a National Green Specification database of all environmentally friendly building materials. This provides a checklist where it is possible to compare specifications of sustainable products to traditionally manufactured products, such as bricks.

Simple changes can be made, such as using timber or ethylene-based plastics instead of UPVC window frames, to ensure a building uses more sustainable materials.

As we have seen, finding locally managed resources such as timber makes sense in terms of cost and in terms of protecting the environment. There are many alternatives to traditional resources that could help protect the environment.

The Timber Trade Federation produces a Timber Certification System. This ensures that wood products are labelled to show that they are produced in sustainable forests.

Around 80 per cent of all the softwood used in construction comes from Scandinavia or Russia. Another 15 per cent comes from the rest of Europe, or even North America. The remaining 5 per cent comes from tropical countries, and is usually sourced from sustainable forests.

Figure 3.27 Window frames made from timber

DID YOU KNOW?

www.recycledproducts. org.uk has a long list of recycled surfacing products, such as tiles, recycled wood and paving and details of local suppliers.

Alternative methods of building

The most common type of construction is, of course, brick and block work. However there are plenty of other options:

* Timber frame

* Insulated concrete formwork – where a polystyrene mould is filled with reinforced concrete.

* Structural insulated panels – where buildings are made up of rigid building boards, rather like huge sandwiches.

* Modular construction – this uses similar materials and techniques to standard construction, but the units are built off site and transported ready-constructed to their location.

Figure 3.28 Timber Certification System

Figure 3.29 Green roofing

Figure 3.30 Flooring made from cork

Alternatives to roofing and flooring

There are alternatives to traditional flooring and roofing, all of which are greener and more sustainable. Green roofing has become an increasing trend in recent years. Metal roofs made of steel, aluminium or copper are lightweight and often use a high percentage of recycled metal. Solar roof shingles, or solar roof laminates, while expensive, help to reduce the use of electricity and heating of the dwelling. Some buildings even have a green roof, which consists of a waterproof membrane, a growing medium and plants such as grass or sedum.

Just as roofs are becoming greener, so too are the options for flooring. The use of bamboo, eucalyptus or cork is becoming more common. A new version of linoleum has been developed with **biodegradable**, **organic** ingredients. Some buildings are also using concrete rather than traditional timber floorboards and joists. The concrete can be coloured, stained or patterned.

An increasing trend has been for what is known as off-site manufacture (OSM). European businesses, particularly those in Germany, have built over 100,000 houses. The entire house is manufactured in a factory and then assembled on site. Walls, floors, roofs, windows and doors with built-in electrics and plumbing, all arrive on a lorry. Some manufacturers even offer completely finished dwellings, including carpets and curtains. Many of these modular buildings are actually designed to be far more energy efficient than traditional brick and block constructions. Many come ready fitted with heat pumps, solar panels and triple-glazed windows.

KEY TERMS

Biodegradable

– the material will more easily break down when it is no longer needed. This breaking down process is done by micro-organisms.

Organic

– this is a natural substance, usually extracted from plants.

Energy efficiency and incorporating it into construction projects

Energy efficiency involves using less energy to provide the same level of output. The plan is to try to cut the world's energy needs by 30 per cent before 2050. This means producing more energy efficient buildings. It also means using energy efficient methods to produce materials and resources needed to construct buildings.

Building Regulations

In terms of energy conservation, the most important UK law is the Building Regulations 2010, particularly Part L. The Building Regulations:

* list the minimum efficiency requirements

* provide guidance on compliance, the main testing methods, installation and control

* cover both new dwellings and existing dwellings.

A key part of the regulations is the Standard Assessment Procedure (SAP), which measures or estimates the energy efficiency performance of buildings.

Local planning authorities also now require that all new developments generate at least 10 per cent of their energy from renewable sources. This means that each new project has to be assessed one at a time.

Energy conservation

By law, each local authority is required to reduce carbon dioxide emissions and to encourage the conservation of energy. This means that everyone has a responsibility in some way to conserve energy:

* Clients, along with building designers, are required to include energy efficient technology in the build.

* Contractors and sub-contractors have to follow these design guidelines. They also need to play a role in conserving energy and resources when actually working on site.

* Suppliers of products are required by law to provide information on energy consumption.

In addition, new energy efficiency schemes and building regulations cover the energy performance of buildings. Each new build is required to have an Energy Performance Certificate. This rates a building's energy efficiency from A (which is very efficient) to G (which is least efficient).

Some building designers have also begun to adopt other voluntary ways of attempting to protect the environment. These include BREEAM, which is an environmental assessment method, and the Code for Sustainable Homes, which is a certification of sustainability for new builds.

energy®
saving
trust

Figure 3.31 The Energy Saving Trust encourages builders to use less wasteful building techniques and more energy efficient construction

High, low and zero carbon

When we look at energy sources, we consider their environmental impact in terms of how much carbon dioxide they release. Accordingly, energy sources can be split into three different groups:

* high carbon – those that release a lot of carbon dioxide

* low carbon – those that release some carbon dioxide

* zero carbon – those that do not release any carbon dioxide.

Some examples of high carbon, low carbon and zero carbon energy sources are given in Table 3.7 below.

High carbon energy source	Description
Natural gas or LPG	Piped natural gas or liquid petroleum gas stored in bottles
Fuel oils	Domestic fuel oil, such as diesel
Solid fuels	Coal, coke and peat
Electricity	Generated from non-renewable sources, such as coal-fired power stations
Low carbon energy source	
Solar thermal	Panels used to capture energy from the sun to heat water
Solid fuel	Biomass such as logs, wood chips and pellets
Hydrogen fuel cells	Converts chemical energy into electrical energy
Heat pumps	Devices that convert low temperature heat into higher temperature heat
Combined heat and power (CHP)	Generates electricity as well as heat for water and space heating
Combined cooling, heat and power (CCHP)	A variation on CHP that also provides a basic air conditioning system
Zero carbon energy	
Electricity/wind	Uses natural wind resources to generate electrical energy
Electricity/tidal	Uses wave power to generate electrical energy
Hydroelectric	Uses the natural flow of rivers and streams to generate electrical energy
Solar photovoltaic	Uses solar cells to convert light energy from the sun into electricity

Table 3.7 High, low and zero carbon energy sources

It is important to try to conserve non-renewable energy so that there will be sufficient fuel for the future. The fuel has to last as long as is necessary to completely replace it with renewable sources, such as wind or solar energy.

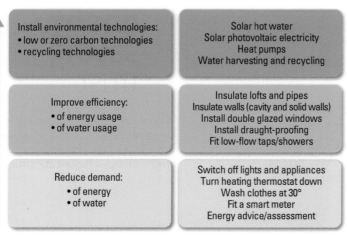

Figure 3.32 Working towards reducing carbon emissions

Alternative energy sources

There are several new ways in which we can harness the power of water, the sun and the wind to provide us with new heating sources. All of these systems are considered to be far more energy efficient than traditional heating systems, which rely on gas, oil, electricity or other fossil fuels.

Solar thermal

At the heart of this system is the solar collector, which is often referred to as a solar panel. The idea is that the collector absorbs energy from the sun, which is then converted into heat. This heat is then applied to the system's heat transfer fluid.

The system uses a differential temperature controller (DTC) that controls the system's circulating pump when solar energy is available and there is a demand for water to be heated.

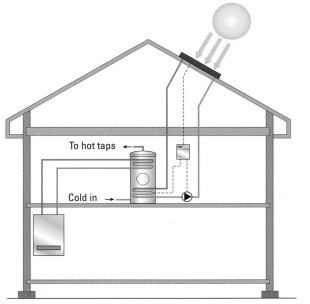

Figure 3.33 Solar thermal hot water system

In the UK, due to the lack of guaranteed solar energy, solar thermal hot water systems often have an auxiliary heat source, such as an immersion heater.

Biomass (solid fuel)

Biomass stoves burn either pellets or logs. Some have integrated hoppers that transfer pellets to the burner. Biomass boilers are available for pellets, woodchips or logs. Most of them have automated systems to clean the heat exchanger surfaces. They can provide heat for domestic hot water and space heating.

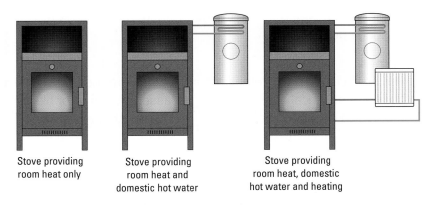

Stove providing room heat only

Stove providing room heat and domestic hot water

Stove providing room heat, domestic hot water and heating

Figure 3.34 Biomass stoves output options

Heat pumps

Heat pumps convert low temperature heat from air, ground or water sources to higher temperature heat. They can be used in ducted air or piped water **heat sink** systems.

KEY TERMS

Heat sink

– this is a heat exchanger that transfers heat from one source into a fluid, such as in refrigeration, air-conditioning or the radiator in a car.

There are different arrangements for each of the three main systems:

* Air source pumps operate at temperatures down to minus 20°C.

* Ground source pumps operate on **geothermal** ground heat.

* Water source systems can be used where there is a suitable water source, such as a pond or lake.

The heat pump system's efficiency relies on the temperature difference between the heat source and the heat sink.

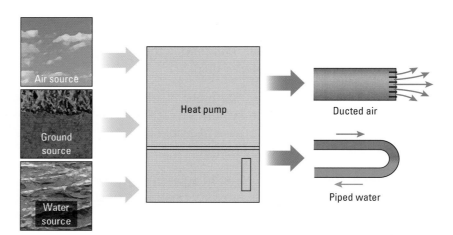

Figure 3.35 Heat pump input and output options

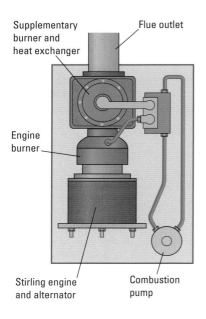

Figure 3.36 Example of a MCHP (micro combined heat and power) unit

Combined heat and power (CHP) and combined cooling heat and power (CCHP) units

These are similar to heating system boilers, but they generate electricity as well as heat for hot water or space heating (or cooling). Electricity is generated along with sufficient energy to heat water and to provide space heating.

Wind turbines

Freestanding or building-mounted wind turbines capture the energy from wind to generate electrical energy. The wind passes across rotor blades of a turbine, which causes the hub to turn. The hub is connected by a shaft to a gearbox. This increases the speed of rotation. A high speed shaft is then connected to a generator that produces the electricity.

Solar photovoltaic systems

A solar photovoltaic system uses solar cells to convert light energy from the sun into electricity.

Energy ratings

Energy rating tables are used to measure the overall efficiency of a dwelling, with rating A being the most energy efficient and rating G the least energy efficient (see Fig 3.41).

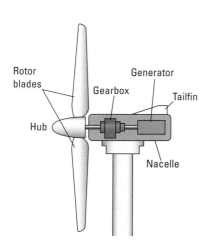

Figure 3.37 A basic horizontal axis wind turbine

Alongside this, there are environmental impact ratings (see Fig 3.42). This type of rating measures the dwelling's impact on the environment in terms of how much carbon dioxide it produces. Again, rating A is the highest, showing it has the least impact on the environment, and rating G is the lowest.

A Standard Assessment Procedure (SAP) is used to place the dwelling on the energy rating table. The ratings are used by local authorities and other groups to assess the energy efficiency of new and old housing and must be provided when houses are sold.

Preventing heat loss

Most old buildings are under-insulated and benefit from additional insulation, whether this is applied to ceilings, walls or floors.

The measurement of heat loss in a building is known as the U Value. It measures how well parts of the building transfer heat. Low U Values represent high levels of insulation. U Values are becoming more important as they form the basis of energy and carbon reduction standards.

By 2016 all new housing is expected to be Net Zero Carbon. This means that the building should not be contributing to climate change.

Many of the guidelines are now part of Building Regulations (Part L). They cover:

* insulation requirements

* openings, such as doors and windows

* solar heating and other heating

* ventilation and air-conditioning

* space heating controls

* lighting efficiency

* air tightness.

Building design

UK homes spend £2.4bn every year just on lighting. One of the ways of tackling this cost is to use energy saving lights, but also to maximise natural lighting. For the construction industry this means:

* increased window size

* orientating window angles to make the most of sunlight – south facing windows maximise sunlight in winter and limit overheating in the summer

* window design – with a variety of different types of opening to allow ventilation.

Solar tubes are another way of increasing light. These are small domes on the roof, which collect sunlight and then direct it through a tube (which is reflective). It is then directed through a diffuser in the ceiling to spread light into the room.

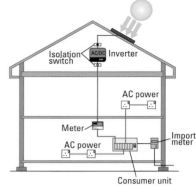

Figure 3.38 A basic solar photovoltaic system

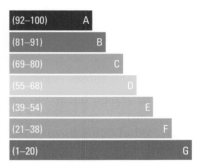

Figure 3.39 SAP energy efficiency rating table – the ranges in brackets show the percentage energy efficiency for each banding.

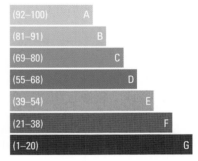

Figure 3.40 SAP environmental impact rating table

TEST YOURSELF

1. In which of the following types of building is a traditional strip foundation used?

   a. High rise

   b. Medium rise

   c. Low rise

   d. Industrial buildings

2. Which of the following is a reason for using a raft foundation?

   a. The subsoil is rock

   b. The subsoil is unstable

   c. The subsoil is stable

   d. The access to the site allows it

3. What holds down a floating floor?

   a. Nails and screws

   b. Adhesives

   c. Blocks

   d. Its own weight

4. What is another term for formwork?

   a. Shuttering

   b. Cavity

   c. Joist

   d. Boarding

5. What is the minimum distance the DPC should be above ground level?

   a. 50 mm

   b. 100 mm

   c. 150 mm

   d. 200 mm

6. A roof is said to be flat if it has a slope of less than how many degrees?

   a. 5

   b. 10

   c. 15

   d. 20

7. What shape is the upper part of a gable end?

   a. Rectangular

   b. Semi-circular

   c. Square

   d. Triangular

8. What do you call the horizontal timber that is placed at the top of a wall at eaves level in a roof, to hold the ends of joists or rafters?

   a. Fascia

   b. Bracings

   c. Wall plate

   d. Batten

9. What happens to the majority of construction demolition and excavation waste?

   a. It is buried on site

   b. It is burned

   c. It goes into landfill

   d. It is recycled

10. Which part of the Building Regulations 2010 requires construction to consider and use energy efficiently?

    a. Part B

    b. Part D

    c. Part K

    d. Part L

Chapter 4

Unit CSA–L1Occ11
PREPARE AND USE CARPENTRY AND JOINERY PORTABLE POWER TOOLS

LEARNING OUTCOMES

LO1/2: Know how to and be able to prepare carpentry and joinery portable power tools

LO3/4: Know how to and be able to use carpentry and joinery portable power tools to cut, shape and finish

LO5/6: Know how to and be able to use carpentry and joinery portable power tools to drill and insert fastenings

INTRODUCTION

The aims of this unit are to:

* help you to maintain and store portable power tools

* show you how to use a range of carpentry and joinery portable power tools.

PREPARING CARPENTRY AND JOINERY PORTABLE POWER TOOLS

Portable power tools can speed up many routine tasks but you must be aware of the risks of using them. According to the HSE, a quarter of all reportable electrical accidents involve portable power equipment.

Hazards and risk assessment

You should always make sure that all necessary precautions have been taken before using a power tool. Many of the points in the following list will be covered in more detail later, but it is necessary to stress their importance here:

* You must make sure that you have had training before you use any tool.

* You should make yourself familiar with the way the tool works and have practised using it in controlled conditions.

* You should give the tool a look over to make sure it is safe to use and know what to do if there is something wrong with it.

* You need to check the plug and cable for any damage.

* If you are using a mains power source, check that it's the correct voltage for the tool, and appropriate to the location of its use.

* If the power tool should have guards you should check that these are correctly fitted.

* Check any blades, drills, cutters or bits before using them. Make sure they are not worn or damaged.

* Make sure you are wearing the right PPE. This could include ear defenders, gloves, a dust mask and eye protection.

* Make sure you are not wearing any loose clothing, as it could easily become caught and injure you or make the power tool overheat.

* Never change a blade, bit, tip or drill without disconnecting the tool from the power supply first.

Risk assessment

Most risk assessments will focus on the tool itself and the electricity supply. Table 4.1 identifies the particular risks and suggests control measures.

Source of risk	Type of risk	Control measures
The power tool	1. Hair, jewellery or clothing getting tangled in moving parts 2. Eye injuries from dust or fragments 3. Hand and wrist injuries from jams and binding 4. Hand or arm vibration syndrome	1. Any loose clothing, dangling jewellery or long hair should be kept clear of moving parts 2. Suitable eye protection should be used if there is a risk of eye injury 3. Tools should only be used in accordance with the manufacturer's instructions and tested at regular intervals 4. Tools with the lowest vibration levels should be used and the amount of time individuals use the equipment should be minimised
The power supply	Electric shock	1. All power feeds should comply with European or British Standards and be in good condition 2. Where practical only 110V electrical tools should be used 3. The tools should only be used in well-lit and well-ventilated areas

Table 4.1

Usually risk assessments are graded according to the exposure to hazards. If a tool is graded as being highly unlikely to cause injury then it can be classed as a safe tool to use. However, many tools are graded as having common or even regular levels of hazard. Each of the different tools will also be graded in terms of the injury that they are likely to cause. At one end of the scale tools may produce a trivial or minor injury, perhaps a blister or a graze. At the other end of the scale they are so dangerous if something goes wrong that they could kill.

230 V mains supply

Types of power sources

There are three main types of power sources that can be used to operate portable power tools:

* either 230V or 110V electric (110V is used on construction sites)

* rechargeable battery packs

* compressed air.

Battery power packs and charger

Portable air compressor

Portable electronic generator

Figure 4.1 Power supplies

Types of portable power tools and their uses

There is a wide range of commonly used portable power tools, each of which has a different use. These can be seen in Table 4.2.

Portable power tool	Description and use
Jig saw	The blade cuts on the upward stroke on most machines, but more expensive versions have an action that moves the blade into the material on the upward stroke and away on the downward stroke. This minimises wear and tear on the blade. A number of different blades are available. They tend to be battery or mains operated at either 110 V or 230 V.
Drill	Drills are perhaps the most common type of portable power tool. Good quality drills can perform a variety of different actions and with the right bits, drills and other accessories they can reduce the effort needed and the time taken.
Drills (hammer and SDS rotary)	These can either be mains or battery driven. Drills are usually rated at 230 V, however, if you use a transformer 110 V tools can be used. On UK construction sites, all power tools have to be 110 V by law. • Battery driven drills (or indeed any battery powered tools) are also known as cordless. They work using a rechargeable battery, which typically range from 12 to 36 V. • There are standard rotary drills, as well as rotary or percussion drills, along with hammer drills.
Planer	A planer is used for chamfering, rebating and edging. On site it is often used to trim the edges of sheet material and for door hanging. They tend to be battery or mains operated at either 110 V or 230 V.
Sander	These machines aim to take much of the hard work out of finishing work. Each different type of sander leaves a more or less smooth surface, so it is important to use the right belt or paper for the job and the material being sanded. Belt sanders are good for removing stubborn defects or old paint. Orbital sanders are good for fine finished work. These tend to be battery operated at 110 V.
Router	Routers have a huge number of uses – from trimming and recessing to dovetailing, drilling, moulding, rebating and grooving. They have a router cutter that can be adjusted on a spring-loaded column. They tend to be operated at 110 V.
Screwdrivers	Most modern drills can also perform the functions of a screwdriver. However there are also mains and battery powered screwdrivers, using standard power supplies. Angled screwdrivers can be used where space is limited, to either remove or insert screws.
Powered nailer	These can either be pneumatic, which uses compressed air, or combustion powered, which uses a cartridge filled with a flammable gas. There are also electric powered nail guns, which either use standard mains electrical supply or are cordless and therefore battery operated.

Table 4.2

Figure 4.2 Power sanders

Figure 4.3 Power drills

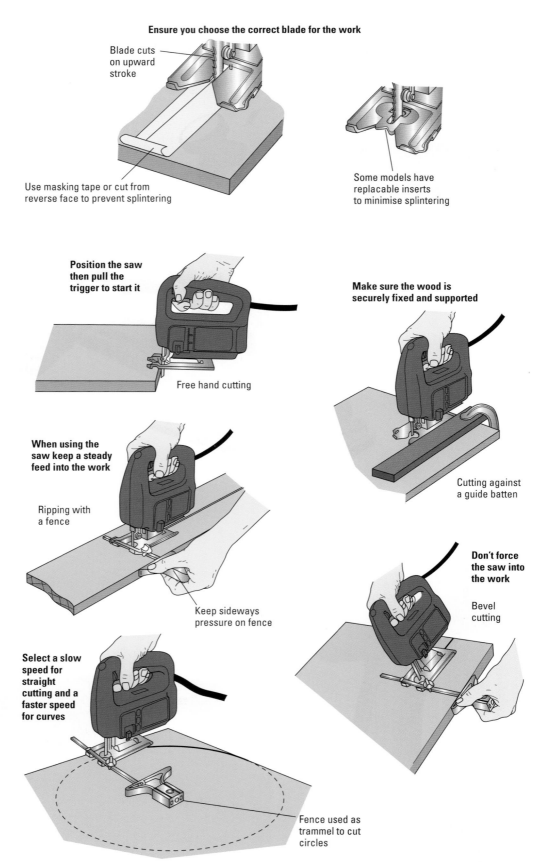

Ensure you choose the correct blade for the work

Blade cuts on upward stroke

Use masking tape or cut from reverse face to prevent splintering

Some models have replacable inserts to minimise splintering

Position the saw then pull the trigger to start it

Free hand cutting

Make sure the wood is securely fixed and supported

Cutting against a guide batten

When using the saw keep a steady feed into the work

Ripping with a fence

Keep sideways pressure on fence

Don't force the saw into the work

Bevel cutting

Select a slow speed for straight cutting and a faster speed for curves

Fence used as trammel to cut circles

Figure 4.5 Operation of a jig saw

No gloves are worn in these pictures, in order to clearly show how to use hand tools. However, you should wear gloves and other PPE required by your college or employer.

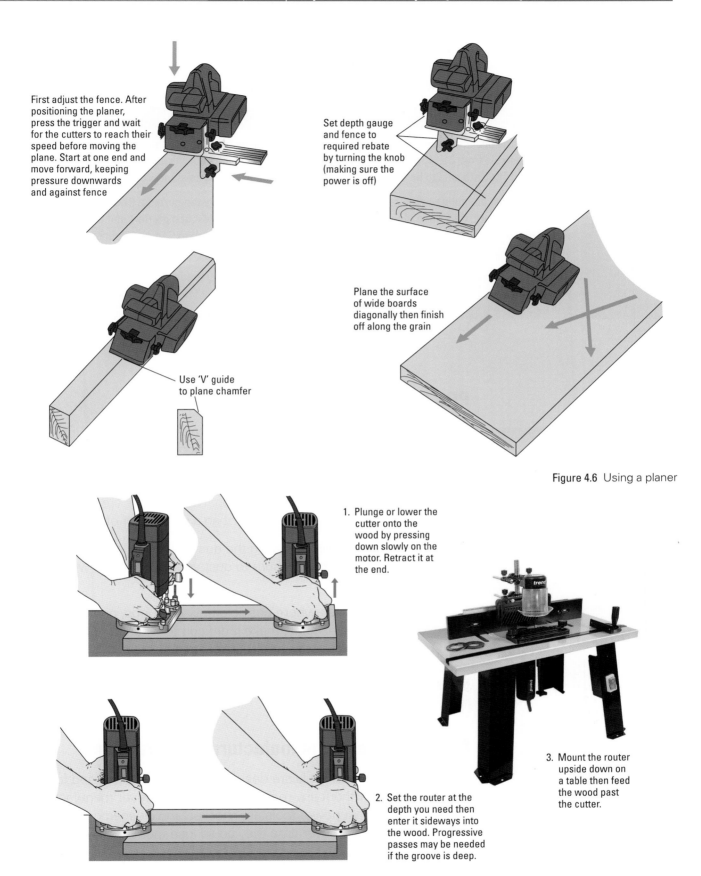

First adjust the fence. After positioning the planer, press the trigger and wait for the cutters to reach their speed before moving the plane. Start at one end and move forward, keeping pressure downwards and against fence

Set depth gauge and fence to required rebate by turning the knob (making sure the power is off)

Use 'V' guide to plane chamfer

Plane the surface of wide boards diagonally then finish off along the grain

Figure 4.6 Using a planer

1. Plunge or lower the cutter onto the wood by pressing down slowly on the motor. Retract it at the end.

2. Set the router at the depth you need then enter it sideways into the wood. Progressive passes may be needed if the groove is deep.

3. Mount the router upside down on a table then feed the wood past the cutter.

Figure 4.7 Three ways of using a router

No gloves are worn in these pictures, in order to clearly show how to use hand tools. However, you should wear gloves and other PPE required by your college or employer.

Drill bits

Table 4.3 shows some of the different types of drill bits that can be selected, together with their main uses.

Type of drill bit	Uses
High-speed steel (HSS)	These are drill bits that have greater resistance to heat and have greater cutting speeds. They are usually available from 2 to 13 mm. They are used for drilling sheet metal or thin materials, including wood and plastic. They are said to be 4 times faster than and twice as strong as standard drill bits.
Polycrystalline diamond (PCD)	These have tiny diamond particles (about 0.5 mm thick) bonded to tungsten carbide. Because diamonds are incredibly hard, they have the ability to drill through even the hardest of materials and are resistant to wear. However, they are brittle and can snap under pressure. As they are extremely expensive they are only used for particular jobs.
Disposable	Disposable bits are cheaper than standard bits because their quality tends to be poorer. However, if you are doing a lot of drilling, it may be more cost effective to spend a little more on standard reusable drill bits.

Table 4.3

PPE

Each different machine and job may require a different type of PPE. In some cases the machines produce a great deal of noise, so ear defenders may be necessary. Any machine that produces dust or particles that could fly up into your eyes, nose or mouth could require you to wear some form of eye protection. Face screens are sometimes more appropriate for better protection than a simple pair of goggles.

To protect yourself against inhaling the dust usually a dust mask is sufficient. For prolonged exposure to dust you may need to wear a respirator.

You should wear gloves where possible but some jobs, particularly when handling the machines, make it difficult to wear any kind of hand protection. If you are in doubt, speak to your trainer or supervisor straight away. If you are assisting in, for example, holding or pushing timber into a machine then you should wear gloves.

PRACTICAL TIP

Gloves are not just worn to protect against splinters and other damage. They are vital as many tools cause vibration and gloves will help prevent any long-term damage to your hands.

Maintenance and manufacturers' instructions

The majority of power tools are designed to be able to be used for long periods of time without any real maintenance. However manufacturers do recommend that power tools are checked and cleaned on a regular basis. This is because over time they could lose their efficiency.

If you are using tools that produce dust or you are working in dusty conditions it is a good idea to clean the tools daily. The dust will be attracted to the motor of the power tool and this could damage it and make it overheat.

Each manufacturer of tools provides an instruction manual. For the majority of situations this will:

* show you how to clean and maintain the power tool yourself. This will tell you what you should do on your own.

* instruct when you should get an authorised repairer to deal with the problem.

You should regularly check your power tools. This means checking for damage on the tool itself and clearing away any dust from ventilation slots. You also need to check the tool's lead and plug for any damage.

Manufacturers will also recommend how to store the power tools. In many cases this means putting them back into their case and making sure that they are not stored in wet or cold conditions.

PRACTICAL TIP

Don't be tempted to unscrew and take to pieces any power tools or have anyone other than an authorised repair person carry out the work. Not only are you putting yourself at risk with a poor repair, but it will also mean that any manufacturer's guarantee on the tool will be invalid.

Legislation

You should always remember health and safety legislation and take precautions when using electricity. In addition to this the following legislation also has to be considered.

Provision and Use of Work Equipment Regulations 1998 (PUWER)
These regulations aim to make sure that any machine used is safe and that it is only used for the right job. You should be trained before you use the machine and you should only ever use it if it has suitable safety measures.

The regulations cover nearly all types of equipment that you might use either in the workshop or on site. This means everything from hammers through to dumper trucks.

The most important thing to remember is that all equipment needs to be suitable for what you are using it for. It needs to be maintained and regularly inspected.

Personal Protective Equipment at Work Regulations 1992
This is a requirement that PPE is supplied and used at work if risks to health and safety cannot be controlled in another way. It covers everything from safety helmets to high visibility clothing.

The regulations state that:

* the most suitable PPE needs to be chosen on the basis that it will provide sufficient protection

* it is stored and maintained in a proper way

* you are given instructions on how to use it properly

* you actually use it in the right way.

Portable Appliance Testing (PAT)

These are visual inspections and electronic tests that check if the appliance is safe and suitable for use. It is the responsibility of the employer to make sure that appliances are inspected and tested on a regular basis. Only people that have been trained and qualified in PAT inspection can carry out these tests. Tools should not be used if they fail a PAT test.

Checking for faults and defects

As we have seen, PAT tests can be used as a regular way of checking the electrical safety of power tools. In the section on maintaining power tools we also discussed the fact that you should never use power tools that are damaged or have damaged parts. If the manufacturer's instructions say that you can replace a part then it should be fairly straightforward and safe to do this. However, you need to make sure you know what you are doing before you try it. If you are unclear you should ask someone more experienced.

If the machine is damaged in any way and you cannot repair it then you must make sure that an authorised repair person deals with the problem.

In addition to checking the tool for faults or defects before you use it, the following are examples of good practice:

Figure 4.8 Electrical PAT safety test

* There should be a regular maintenance programme and a maintenance log kept. This might mean daily, weekly or monthly checks.

* At least every 12 months a competent person should carry out a PAT test and records of these inspections and tests should be kept. The frequency of PAT tests is dependent on the tool's risk factor and how much it is used.

Preparing carpentry and joinery portable power tools

The following practical exercise looks at how to prepare power sources for power tools. It also covers how to check the tools, cables and tooling, as well as how to change tooling following the manufacturer's instructions.

PRACTICAL TASK

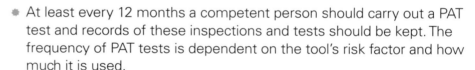

1. MAINTAIN CARPENTRY AND JOINERY POWER TOOLS

OBJECTIVE

To know how to maintain portable powered machinery.

For tasks on using and storing power tools, refer to the practical tasks from p210

The tools used could be electrically powered, battery powered or air powered.

ROUTER CHECKLIST

* Check the power supply.
* Check the power cord is in good condition with no fraying or cracks.
* Make sure the router has been PAT tested.
* Check the overall condition of the router: is it in good condition?
* Check that all stops and guides are functioning correctly.
* Make sure that you are fully trained in the use of such machines.

MAINS RADIAL ARM SAW CHECKLIST

* Check the power supply.
* Check the power cord is in good condition with no fraying or cracks.
* Make sure the saw has been PAT tested.
* Check the overall condition of the saw: is the guarding in place?
* Check the condition of the blade and ensure it is the correct blade for the machine.
* Check that holding down cramps are in position.
* Make sure that you have been trained in the use of the particular saw.

JIG SAW CHECKLIST

* Check the power supply.
* Check the power cord is in good condition with no fraying or cracks.
* Make sure the jig saw has been PAT tested.
* Check the overall condition of the jig saw.
* Check that the correct blade is properly fitted.
* Check that the base is secure and square to the blade.
* Make sure that you are fully trained in the use of such machines.
* Check that the speed and reciprocating action are set correctly.

ORBITAL SANDER CHECKLIST

* Check the power supply.
* Check the power cord is in good condition with no fraying or cracks.
* Make sure the sander has been PAT tested, if applicable.
* Check the overall condition of the sander.
* Check that the base is in good condition with no hollows etc.
* Check that the holding down cramps for the sanding sheets are working correctly.
* Ensure the sanding sheet is securely held in the cramps and that it is flat to the base.
* Make sure that you are fully trained in the use of such machines.
* Check that the speed and reciprocating action are set correctly.

PORTABLE POWERED CIRCULAR SAW CHECKLIST

* Check the power supply.
* Check the power cord is in good condition with no fraying or cracks.
* Make sure the circular saw has been PAT tested, if applicable.
* Check the overall condition of the circular saw.
* Check the condition of the blade and that it is the correct type.
* Check that the guarding is operating correctly.
* Make sure that you are fully trained in the use of such machines.

USING CARPENTRY AND JOINERY PORTABLE POWER TOOLS TO CUT, SHAPE AND FINISH

At the end of this section a practical exercise covers how to select and use the right tooling for different jobs. It also looks at selecting the right PPE, how to use holding devices, such as vices and cramps, and then how to go about cutting, shaping and sanding different kinds of timber.

In this first part of the section we look at the fact that tooling such as blades, bits and tips can be damaged over time. We also discuss the fact that many of these machines produce debris, which in itself is hazardous.

Tooling and damage

Regardless of the cost of the tip, blade or bit the tooling will eventually have to be changed. It can wear, fracture or become rough and far less accurate.

There is not always a good way of knowing how long the tooling will last. A chip can gradually become more serious and this may depend on the roughness or the strength of the material that you are cutting, moulding, shaping or sanding.

Damaged blades, for example, will reduce the cutting speed and probably make the cut far rougher than it should be. Worn drill bits will require a higher speed and more pressure to drill holes into material.

If you buy good quality drill bits and make sure that you use the right speed and lubricate them, they can last without re-sharpening for a long time. By slowing down the speed of the drill it might take a little longer to drill the hole, but you will not have to change the bits so often. It is the speed of the drill that causes the heat and it is the heat that causes the damage to the drill bit.

Most jig saws will come with standard blades. These might work well for simple DIY projects, but at work you will need a wider range of different jig saw blades for different jobs. They are classified by how many teeth they have per inch. Always make sure that you choose the right blade for the right material. Jig saw blades are thin and are only really supported at the top, where they are attached to the jig saw. This means that they can bend and overheat quite easily. Bending is more likely to happen if you are cutting through timber that is too thick.

High-speed jig saw blades do not actually move any faster than normal blades but they do have greater penetration and can be used on thin metals. You should only ever use cobalt steel blades for thicker wood and metals. These blades can be sharpened several times before you need to replace them.

Many manufactured boards, such as plywood, include glue in their manufacture. These can reduce the life of tooling. The glue is very hard compared to the rest of the material and it dulls blades and drill bits.

PRACTICAL TIP

Remember that when the tools are working on materials they will heat up. Some of them are very sensitive to temperature changes. This makes the wear and tear on the tool speed up.

PRACTICAL TIP

All drill bits will eventually become blunt and will need replacing. Sanding belts get clogged up with sawdust, which affects the finish of the timber and effectiveness of the tool. So these also need replacing regularly.

DID YOU KNOW?

Tungsten carbide-tipped blades will stay sharper for longer. But they cannot be mended if they are damaged. High-speed steel blades can be re-sharpened.

DID YOU KNOW?

Carbide blades can actually cut through masonry board and porous concrete. They will need to be sharpened and replaced on a regular basis. These are much harder and more heat resistant than most other types of blade.

If you are drilling or cutting through hardwood, such as oak, then tooling can get dull very quickly. This is also true if the wood has any knots.

Lubricants have a wide variety of different uses. There are a number of traditional types, such as oil or paraffin, but increasingly plant oil is being used. The key benefits of using lubricants are:

* they are rust inhibitors (they stop rust)

* the blade is cleaner

* the times between having to sharpen the blades are longer

* the blade tension is retained for longer.

Debris, hazards and the importance of a clean working area

Efficient dust collection is vital for health reasons but also to comply with the law. Power tool users can suffer from allergic reactions, which can affect the nose, eyes and skin. A build-up of dust can also pose a fire hazard. Larger particles that cling to surfaces can cause scoring and there is also the problem of reduced visibility, which can make accurate measurement and cuts impossible.

A simple dust collection system uses a duct system. This moves the dust from the saw to a collection device that is attached to the ducting. Metal ducting is usually thought to be better than plastic piping. This is for three reasons:

* There is a limited choice of suitable plastic pipe fittings that would meet the needs of the extraction.

* The elbows in plastic pipes tend to clog.

* Plastic piping is **non-conductive** – it builds up a static charge as the charge particles pass along it. This charge can shock and there is also the risk of explosion or fire.

Spiral, steel pipe with fittings that have a long radius are less likely to clog. They can also be fitted with sections that can unclip and be cleaned out. The pipe is **conductive** and is less likely to be a fire hazard.

Nearly every task will produce some waste, whether it is dust or small pieces of wood. The Building Act (1984) clearly states that it is the construction industry's responsibility to prevent and control waste. It should also make sure that resources are not wasted unnecessarily.

The Building Regulations cover the problem of waste disposal. This is in Part H, which covers all types of building materials, including wood.

The drive towards sustainable and secure buildings also aims to control waste and to protect the environment. You could refer back to Chapter 3, to refresh your memory about sustainability.

In order to reduce the amount of waste Table 4.5, which follows, can be used as a guide.

KEY TERMS

Non-conductive

– this is a material that does not readily conduct electricity, and static electricity may build up in the material.

Conductive

– this means that an electrical current can pass through the material and will not build up in it.

Waste reduction and disposal method	Explanation
Elimination	This involves not producing the waste in the first place. Regularly checking stocks of materials on site or in a workshop stops over-ordering. Using cutting lists means you can order the right lengths of materials and reduce the waste. Once the materials have arrived, if they are stored properly they will be in a good state for another job. It is also important to cut as accurately as you can to minimise accidental offcuts.
Reduction	Always keep materials in their protective packaging and try not to handle the material unless necessary as this will avoid damage. Always put materials back into storage. If you have large offcuts set them aside as they might be useful later. Always use up opened stock before breaking into a new package.
Re-use	Use offcuts for pegs, profile boards and repairs. Re-use timber offcuts as many times as you can. They can be used for hoardings or form work.
Recycle	Most timber can be recycled. Some has a high value, such as reclaimed oak or pine for furniture. Most other timber, no matter how small the offcut, can be used to produce chipboard or MDF. You should throw wood into a skip only as a last resort.

Table 4.4

Cutting, shaping and sanding

This practical task looks at how to select and use the right tooling for manufacturing a two panelled door, including the fitting of a mortise lock and night latch, using the following power tools:

* Radial arm saw
* Portable powered planer
* Circular saw
* Portable router
* Drill
* 110V transformer
* Jig saw
* Orbital sander

It would not be normal practice to produce a door with these tools; however, this is an example of the versatility of this machinery and some of the ways they can be used. For this reason, no plans or measurements are given for these tasks. Each different job may need a different holding device, such as a vice, cramp or jig.

PRACTICAL TASK

2. CUT, SHAPE AND SAND TIMBER AND MANUFACTURED BOARD

OBJECTIVE

The objective for this and the tasks that follow is to use and store different power tools

Perform your pre-use checks for the radial arm saw and circular saw, as shown in the previous task.

PPE

In this and the tasks that follow, ensure you select PPE appropriate to the job and site where you are working. For all practical tasks in this chapter, refer to the PPE section in Chapter 1. No gloves are worn in these pictures, in order to clearly show how to use hand tools. However, you should wear gloves and other PPE required by your college or employer.

STEP 1 Set out the door from the rod. Mark the position of all mortise and tenons. Remember rebated stiles and rails will require long and short shoulders. See *Setting out from rods* in Chapter 6.

STEP 2 Plug the radial arm saw into the mains supply via a 110 volt transformer.

STEP 3 Using PSE timber of stock sizes, cut to the required lengths, allow for horns on the stiles and some additional length on the rails, using a chop or radial arm saw or circular saw.

STEP 4 The saw should be cleaned down after use and a visual check should be made to ensure the saw has not sustained any damage during use. The saw and transformer should be stored in a dry, secure store with the cable properly coiled.

PRACTICAL TASK

3. FORMING THE MORTISE HOLES USING A PORTABLE POWERED ROUTER

To form the mortises, a powerful router that accepts half-inch cutters will be required.

* Perform your pre-use checks for the router.

STEP 4 Plug the router into the mains supply via a 110 volt transformer.

STEP 1 Make a simple jig that restricts the movement of the router to the outer edges of the required mortise.

Figure 4.9 Making a simple jig with a jig saw

STEP 2 Ensure the router is fitted with a cutter that has a diameter equal to the mortise width and will reach just over half the width of the stiles.

STEP 3 Cramp the stiles securely to a bench or suitable work surface.

STEP 5 Plunge the router to a depth of approximately 10 mm and make the first pass, continue to increase the depth with each pass until the depth is just below half way.

Figure 4.10 Using the router

STEP 6 Turn the stiles over and repeat until the mortises are completely through the stiles.

STEP 7 Form the slots for the haunches to the depth required. Set the depth stop on the router to make sure all haunches are equal. Always allow the cutter to stop before removing from the piece.

PRACTICAL TIP

It may be necessary to clamp additional pieces of timber of the same section either side of the piece to offer a better bearing for the base of the router.

STEP 8 Square up the ends of the mortises with a sharp chisel.

STEP 9 Clean the router after use and make a visual check to ensure the router is not damaged.

Store the router and transformer in a purpose-made box or carrying case, usually supplied with them. Always ensure that the cable is properly coiled. Place in a dry secure store.

PRACTICAL TASK

4. FORMING THE REBATES USING A PLANER

STEP 1 Set the depth stop to the depth of the required rebate. The depth stop is located on the side of the planer and is usually secured with a wing nut or similar.

STEP 2 Fit the rebating fence to the planer. This is usually connected by sliding bars that slot into the body of the planer. Set the fence to the width of the rebate.

Figure 4.11 Fitting the rebating fence

STEP 3 Set the depth of cut by adjusting the in-feed bed on the sole of the planer.

Figure 4.12 Adjusting the in-feed bed

STEP 4 Plug the planer into the mains supply via a 110 volt transformer.

STEP 5 Make several passes until the depth stop bottoms out on the surface of the stile or rail. Complete all rebating before moving on to the forming of the tenons.

Figure 4.13 Using the planer

PRACTICAL TIP

Always allow the planer to stop before removing from the piece.

STEP 6 Clean down the planer after use and give it a visual check to ensure it has not been damaged. Pay particular attention to the blades: are there any chips, are they fully tightened or is there a build of resin?

The planer and transformer should be stored in a purpose-made box or the carrying case supplied with them, with the cable properly coiled. This in turn should be kept in a dry secure store.

PRACTICAL TASK

5. FORMING THE TENONS USING A RADIAL ARM SAW

Perform your pre-use checks for the radial arm saw.

Some radial arm saws have adjustable stops that allow the saw to be pulled across the timber at a given depth. This allows the saw to be pulled across the piece to form the shoulders. By moving the piece along in small increments, tenons can be formed.

STEP 1 Mark the cheeks of the tenons with a mortise gauge.

STEP 2 Lay the timber face side down and set the depth stop on the saw.

STEP 3 Make sure the rails are clamped down using the cramps provided on the saw.

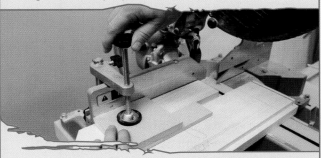

Figure 4.14 Clamping the rails

STEP 4 Cut the shoulders of the tenons on both sides. Set up a stop so all the shoulders are cut in line. Remember the shoulders will be long and short to allow for the rebate.

Figure 4.15 Cutting the shoulders

STEP 5 Make a series of passes over the timber to form the tenon. Clean up the cheeks with a badger or shoulder plane. Repeat for all tenons.

Figure 4.16 Forming the cheeks

Figure 4.17 Cleaning up the cheeks

PRACTICAL TIP

Alternatively tenons can be quite easily formed with the use of a router and a basic jig. Once the shoulders have been formed all that remains is for the waste to be removed with a series of passes.

STEP 6 The saw and transformer should be stored as described for the previous tasks.

6. CUTTING THE HAUNCHES USING A JIG SAW

Perform your pre-use checks for the jig saw.

STEP 1 Mark out the haunches onto the cheeks of the tenons.

STEP 2 Plug the jig saw into the mains supply via a 110 volt transformer.

STEP 3 Clamp the stiles to a suitable work surface.

STEP 4 Using a spacing piece to sit on the cheek of the tenon, cut out the haunches on each of the stiles.

Figure 4.18 Cutting the haunches

STEP 5 Before putting the jig saw away cut enough wedges from a piece of scrap timber for each tenon on the frame.

Figure 4.19 Wedges cut from scrap

PRACTICAL TIP

Always allow the saw blade to stop before removing from the piece.

STEP 6 Clean the jig saw down after use and make a visual check that it has not sustained any damage during use.

Store the jig saw and transformer in a purpose made box or carrying case, usually supplied, with the cable properly coiled. This in turn should be kept in a dry secure store.

7. ASSEMBLING THE FRAME

STEP 1 Dry fit the frame and place in sash cramps, check the fit of all joints and check for square.

Figure 4.20 Dry fitting the frame

STEP 2 Take the door out of the cramps and chop wedge room on all mortises using a sharp mortise chisel of the correct width.

Figure 4.21 Chopping wedge room

STEP 3 Apply glue to both cheeks of each tenon and along the shoulders, and assemble the door.

STEP 4 Place in the sash cramps and square as before.

STEP 5 Apply glue to each wedge and insert using a hammer.

Figure 4.22 Inserting the wedges

STEP 6 Remove from the cramps.

STEP 7 Allow glue to set.

PRACTICAL TASK

8. CLEANING UP USING AN ORBITAL SANDER

Perform your pre-use checks for the orbital sander.

PRACTICAL TIP

Some orbital sanders do not use cramps to secure the sanding sheets to the base of the machine. Instead they use self-adhering sheets. These are matched to the particular make and model and are more expensive than buying rolls of sandpaper.

STEP 1 Ensure the door is on a flat surface and held securely.

STEP 2 Fit a piece of 80 grit sandpaper into the sander.

Figure 4.23 Fitting the sandpaper

STEP 3 Plug the sander into the mains supply via a 110V transformer.

STEP 4 Start with the high points on the door surface; these are most likely to be at the joints. Work along the grain as this will reduce the chances of creating hollows on the surface. Do not work any one area too much, and keep the sander moving.

PRACTICAL TIP

Where more material needs to be removed, this would normally be done with a belt sander.

STEP 5 Now sand each rail in turn until the whole rail has been sanded.

Figure 4.24 Sanding the rails, working along the grain

STEP 6 Repeat for the stiles.

STEP 7 Turn the door over and repeat Steps 3 to 5.

STEP 8 Fit a piece of 120 grit sandpaper into the sander.

STEP 9 Repeat Steps 5 and 6.

STEP 10 Clean down the sander after use and give it a visual check to ensure that it has not sustained any damage during use.

Store the sander and transformer in a purpose made box or the carrying case supplied with them, with the cable properly coiled. This in turn should be kept in a dry secure store.

PRACTICAL TASK

9. CUTTING OFF THE HORNS USING A PORTABLE POWERED CIRCULAR SAW

Perform your pre-use checks for the portable powered circular.

STEP 1 Measure the distance from the edge of the base to the edge of the saw blade (A). This will be different depending on which side of the blade you are working from. If you are cutting from right to left you will be using the wider side of the base, but if you are working from left to right you will be using the narrower side of the base.

Figure 4.25 Measuring the distance to the saw blade

STEP 2 Clamp a temporary guide across the door at the distance (A).

Figure 4.26 Clamping a temporary guide across the door

STEP 3 Set the depth of the cut by moving the base plate. Allow enough blade depth to allow the cut to be made without excessive blade protruding beneath the door.

Figure 4.27 Setting the depth of the cut (1)

Figure 4.28 The blade should not protrude too much beneath the door

STEP 4 Plug the circular saw into the mains supply via a 110V transformer.

STEP 5 Place the saw against the temporary fence and switch it on, making sure both hands are clear of the blade, and make the cut. Turn off the saw and allow the blade to stop before lifting from the work.

STEP 6 Repeat for the horns at the other end of the door.

STEP 7 Clean and store as for the other power tools.

USING CARPENTRY AND JOINERY PORTABLE POWER TOOLS TO DRILL AND INSERT FASTENINGS

Fastening and fixing to timber products can be carried out using a range of different types of screws, bolts, nails and adhesives. Each of the different types can be used for a variety of different surfaces, although some are more appropriate for different surfaces.

Plastic plugs

Plug fixings are used to attach timber to masonry. The plug is inserted into the hole that has been created using an electric drill. In the past carpenters used to make their own plugs out of timber, but nowadays it is far more common to use ready-made plastic plugs.

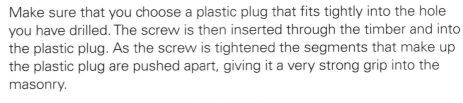

Figure 4.29 Plastic plugs are available in a variety of different sizes

Make sure that you choose a plastic plug that fits tightly into the hole you have drilled. The screw is then inserted through the timber and into the plastic plug. As the screw is tightened the segments that make up the plastic plug are pushed apart, giving it a very strong grip into the masonry.

Rawl bolts/expansion anchor bolts

Rawl bolts are useful if you are drilling into brickwork, stone or concrete. You should avoid drilling into mortar joints. The rawl bolt is basically an expanding, anchored fastener. You will need to drill a hole of the same diameter as the rawl bolt. The nut and washer is removed and the anchor is put into the hole. The timber is then fixed over the thread and then the washer and nut are fastened back onto the bolt.

Coach screws

A coach screw looks like a screw but has the head of a bolt. They are used for a wide variety of timber work in buildings. They are considered to be heavy duty. The shaft tapers to a point and as the screws are driven into the timber, using either a wrench, spanner or pliers, the head stops it from going all the way through. Coach screws are often used at major joints in timber. They are considered to be better than nails, mainly because they are stronger and can be removed easily if necessary.

Figure 4.30 A rawl bolt/expansion anchor bolt

Screws

Screws are graded according to their head type, their length and their gauge or diameter. Carpenters use screws with countersunk heads when the screw needs to be flush or below the surface of the material. Round-headed screws are used when sheet material is attached to timber. Raised head screws are normally used to attach metal fixings, such as door handles.

Screws also have different types of heads and there are screwdrivers or tools to match each type and size. The three common head types are Slotted, Phillips and Pozidriv. You may also come across Torx, Supredrive, Spax and Clutch.

Figure 4.31 A coach screw

Cavity fixings

Different sorts of fixings have been designed to cope with hollow walls.

There are a number of different options:

* Rubber sleeved fixing (Rawlnut) – a hole is drilled in the board and as the screw is tightened the rubber sleeve is compressed against the reverse side of the wall.

* Plastic collapsible fixing (Poly-Toggle) – this is a very similar method to the rubber sleeve, but instead uses a collapsing plastic sleeve. As the screw is tightened the plastic sleeve is pulled back on itself tightly against the inside of the board.

* Plastic spread fixing (plasterboard plug) – the plastic fixing is pushed into the drilled hole and an ordinary wood screw is tightened. The

Figure 4.32 Screws are available in a variety of sizes

fixing is pushed through the hole in the board. Small plastic legs of the fixing are then compressed against the back of the board to hold the screw in place.

* Spring metal toggle – these are stronger anchors and consist of a pair of spring-loaded metal arms with a thread tapped into a hinge pivot. The hole is drilled and the device is then pushed through the hole. Once the spring arms are free of the hole they will spread out onto the reverse of the board. The screw is then tightened, which compresses the arms against the back of the board.

* Gravity metal toggle – these work in a similar way to the spring metal toggle except that the toggle is only made of one piece of metal. It lies parallel to the screw as it is pushed through the drilled hole. Once it is through the hole, gravity takes over and the toggle drops to a vertical position. The screw is then tightened, which compresses the toggle against the back of the board.

Figure 4.33 Cavity wall fixing

Coach bolts

A coach bolt has a domed head with a square shank. A hole is drilled to accept the bolt and is tightened using a nut. They are often used for connecting timber trusses and are considered to be heavy duty and strong fixings.

Nails

Nails can be fixed into timber using either a hammer or a mechanical tool, such as a nail gun (you should never use a nail gun if you have not been trained to do so). Nails are either made from ferrous metal (this means they contain iron and could rust) or non-ferrous metal (which does not contain iron).

Figure 4.34 A coach bolt

There is a huge variety of different nails in terms of style, shape and size:

* Round wire – these are generally for first fixing work, or low quality jobs and tend not to be driven below the surface of the timber so they can easily be removed.

* Annular ring – these have a number of rings along their shank that gives them a stronger hold but at the same time makes them tougher to get out of the timber. They are covered in a thin layer of zinc to reduce the chance of them rusting.

* Lost head – these are nails that are used to fix floor boards. They can be driven in without the need to punch them below the surface due to the shape of the head.

* Oval wire – these are designed to go beneath the surface of the timber and because of their shape they allow the grain to close around the timber as they are punched below the surface.

* Panel pins – these are thin nails that are used to fix fine mouldings and beading. They are circular in cross-section and are designed to be punched below the surface of the timber.

* Cut nails and floor brads – cut nails are made from mild steel and are square in section. They give a good grip. They are used for fixing timber to blockwork. Floor brads are very similar to cut nails and are used for surface fixing of floorboards.

* Plasterboard nails – these are treated to prevent them from rusting and have a rough shank to give them a good grip. They are used to fix plasterboard or insulation board to joists and stud work.

* Wire clout – these are short nails with large heads and are used for fixing roof felt.

* Hardened steel – these are similar to lost head nails but are made of hardened zinc-plated steel. They are used mostly for fixing to brickwork and concrete.

* Duplex – these have double heads. The lower head is driven into the surface of the material and the upper head stays above the surface, making them easier to pull out. They are mainly used for temporary fixings.

* Machine-driven – these are nails and pins that are in coils or strips and used by pneumatic nailing machines. There is a huge range of different types and they can be used for structural work or final fixing of mouldings and trims.

Figure 4.35 Nails are available in many different sizes

Chemical fixing

There is a wide range of different adhesives that create a chemical bond, rather than relying on punching a hole through the material with a nail, screw or other type of fixing.

The adhesive always needs to penetrate the surface and key into the layers. This is a process known as mechanical adhesion.

The adhesive also has to have sufficient strength. This is why adhesives tend to be applied in a liquid state. As they harden, set or cure they become solid and stronger. This is achieved in a number of different ways:

* Solvent adhesive – when the adhesive is applied the solvent evaporates, or is absorbed into the timber.

* Cooling – a hot glue gun heats up the adhesive, transforming it from solid to liquid. It is applied in the liquid form but as it cools it hardens.

* Chemical – the adhesive needs a hardener or another chemical, or sometimes heat, to make it transform from a liquid to a solid. Synthetic resin adhesives are a good example. These are two-part powder or liquid adhesives.

* Combination – some adhesives use their loss of solvent, a chemical reaction and cooling at the same time to make them strong.

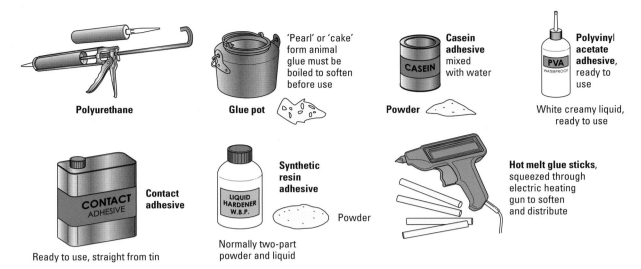

Figure 4.36 Types of adhesive

Safe use of chemical fixings

Always ensure that you read the instructions on the container of each chemical fixing you select for a task. This information will provide you with details of:

* its safe and effective use

* safety equipment required (PPE)

* correct storage and transportation

* what to do if you accidentally swallow or get the product on your skin or in your eyes.

Locating services

Whenever you are drilling or cutting there is a danger that just beneath the surface is either a cable or a pipe. It does not necessarily follow that water or gas pipes will only be found under floorboards. There also may be no logic behind the direction electric cables take inside walls. Not even cavity walls can be considered safe, as the cavity may have been used for electric cables and pipes. This is the same for partition walling inside a building.

There are devices that you can use to detect the flow of water, the presence of metal or an electric current.

PRACTICAL TIP

It is never a good idea to absolutely trust service detectors or locators. If you think you are going to have to drill or cut in an area where there might be services the wise precaution is always to turn those services off at the mains. This means that if any damage is done to pipes or cabling then there will not be an immediate danger and they can be repaired before the mains are switched back on.

TEST YOURSELF

1. What voltage of electrical tools should be used to minimise the risk of electric shock?

   a. 110V

   b. 150V

   c. 230V

   d. 240V

2. Which of the following tasks is made easier by a planer?

   a. Chamfering

   b. Rebating

   c. Edging

   d. All of these

3. What are the particles made of that make PCD drill bits especially strong?

   a. Carbon

   b. Tungsten

   c. Gold

   d. Diamond

4. What is a PAT test?

   a. A competency test to check you can use power tools safely

   b. An inspection made to an appliance that has broken

   c. A test for an appliance's safety before its first use

   d. A test to check if an appliance is still safe to use

5. When dealing with waste, elimination, reduction and re-use are all recommended. But which reduction and disposal method has been missed out?

   a. Landfill

   b. Recycle

   c. Repair

   d. Re-order

6. Which fixing device is useful if you are trying to fix to brickwork, stone or cement?

   a. Rawl bolt/expansion anchor bolt

   b. Coach screw

   c. Cavity fixing

   d. Nail

7. What is the term used to describe nails that have double heads?

   a. Annular ring

   b. Wire clout

   c. Duplex

   d. Machine driven

8. What feature do plasterboard nails have to give them a good grip?

   a. An enlarged tip

   b. A larger head

   c. A thicker diameter

   d. A rough shank

9. What metal is used to make cut nails and floor brads?

   a. Mild steel

   b. Iron

   c. Zinc

   d. Brass

10. Service locating machines can detect which of the following?

    a. Metal

    b. Water flow

    c. Electric current

    d. All of these

Chapter 5

Unit CSA–L2Occ36
CARRY OUT FIRST FIXING OPERATIONS

LEARNING OUTCOMES

LO1/2: Know how to and be able to prepare for first fixing operations

LO3/4: Know how to and be able to install timber frames and linings

LO5/6: Know how to and be able to install timber floor coverings and flat roof decking

LO7/8: Know how to and be able to erect timber stud partitions

LO9/10: Know how to and be able to install straight flights of stairs and handrails

INTRODUCTION

The aims of this chapter are to:

* help you select resources and carry out the work

* help you to erect and fix first fixing components in accordance with the work specification.

FIRST FIXING OPERATIONS

The term first fix is common in the construction industry. It involves all operations that take a building from its foundations through to plastering.

For carpentry work this means involvement in the construction of walls, floors, ceilings and stairs, as well as many other activities. This chapter covers those activities required at Level 2:

* fixing frames and linings

* fitting and fixing floor coverings and flat roof decking

* erecting timber stud partitions

* assembling, erecting and fixing straight flights of stairs including handrails.

Working drawings, specifications and schedules

Drawings and their purpose are covered in detail in Chapter 2.

All drawings must follow the requirements of BS 1192:2007. This means that the drawings will have a common format and symbols. Building drawings use what is known as first angle orthographic projection. Drawings in this form will be identified by a special symbol.

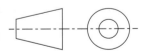

Figure 5.1 First angle projection symbol

Datum points

A site datum point is a fixed point against which all other levels on the site can be measured. It is usually located at the damp-proof course level. Site datums are now usually related to the Ordnance Survey national grid, a standard grid that applies across the country. It is at a known height above sea level. GPS may also be used. Less commonly, the site datum is related to another fixed point, such as:

* an **Ordnance Survey Benchmark (OSBM)** on public buildings or cut into walls, although these are not used as much these days because often they are not maintained

* curb edge

* maintenance/inspection chamber frame

* TBM (temporary benchmark), which could be any agreed item but may change as the site develops

* ABM (arbitrary benchmark), which could be a relatively permanent point, like the plinth of a building, with its level given an arbitrary height that may not be its actual height.

The site datum is marked on the site by a steel post or peg concreted into the ground, usually near the site office.

PPE

Refer to Chapter 1 for details about PPE required for work on site and elsewhere.

Tools and equipment

On site you will use a wide variety of different tools and equipment, depending on the job. You will need to be able to select the correct tool or piece of equipment for each of these jobs. They are detailed in Table 5.1.

Figure 5.2 The OSBM symbol

Tool or equipment	Use
Saws Figure 5.3 Cross-cut saw Figure 5.4 Panel saw Figure 5.5 Tenon saw	There is a wide variety of different saws. Some are modern hard point disposable saws but others are traditional hand saws that can last a lifetime. Saws include: • Cross-cut – for cutting timber across the grain • Panel – for fine cross-cutting, particularly plywood and hardboard. Cross-cut and panel saws are used at an angle of less than 45° to prevent tear out. Their teeth are sharpened at an angle • Tenon – a general purpose bench saw which provides a higher accuracy in cutting, due to its rigid spine along the top of the blade • Rip – for cutting along or with the grain. They are used at approximately 45° with teeth that cut like a chisel. They are sharpened on their front edge • Pull – used for ripping and cross-cutting • General or fine hard point panel saws– for most carpentry activities • Coping – for cutting out curved work and scribing on internal mouldings such as complex skirting profiles.

Tool or equipment	Use
Hammers  Figure 5.6 Claw hammer  Figure 5.7 Club or lump hammer	There are a number of different hammers used and on site these generally include: • Claw – a general all-purpose hammer, for nailing or for taking out nails. They are made of a variety of materials but mainly have a steel head with a steel, wood or fibreglass shaft • Mallet – can either be wedge or round-headed. The wedge shape is more common on site. A rubber mallet is most commonly used when carrying out bench joinery work • Warrington or cross pein – used for lighter work such as nailing glazing beads and mouldings • Lump hammers – used for heavier work such as chopping out masonry.
Chisels  Figure 5.8 Bevel-edged chisels	Split-proof handled chisels that have impact-resistant plastic handles are most favoured on site. Conventional wooden chisels are more likely to be used for bench work. Bevel-edged chisels sit tight into corners of trenches or housings. They are particularly useful when producing dovetail joints. Mortice chisels are used for dropping the mortice half of a mortice and tenon joint. They are very strong and are designed not to snap when chopping deep into the timber. They are rectangular in cross section.
Screwdrivers  Figure 5.9 A selection of screwdrivers	Pozidriv screwdrivers have star-shaped heads. Phillips screwdrivers have cross heads. Slotted screwdrivers are rectangular in section. All are available in different sizes to match screw heads and as stubby versions for use where there is limited access.
Spirit levels  Figure 5.10 Spirit levels	These are used for plumbing and levelling. They can be up to 2 m long. Generally though an 800 mm spirit level will work for most jobs. They are available as either box or beam; beam is the preferred and better quality type of level. Laser versions are available. Smaller levels are available for work in more restricted areas. Boat levels are used for levelling window boards and shorter levelling tasks with greater accuracy than using a longer spirit level.
Plumb bobs  Figure 5.11 Plumb bob	A plumb bob is a weight on the end of a string line used to find plumb. They are usually made of brass and often decoratively turned. They are available in different weights and provide a high level of accuracy over longer distance plumb levelling.

Tool or equipment	Use
Electric and cordless drills  Figure 5.12 Impact drill	Multipurpose drills can perform a variety of tasks. They fall into the following categories: • Rotary impact – these produce a hammer action, ideal for masonry and concrete • Rotary hammer – these go through masonry and concrete very quickly • Electro-pneumatic – much more powerful than normal hammer drills. There are also battery-powered versions. Drills can be slowed down for use as screwdrivers or they can incorporate impact or rotary hammer functions. Power tools on construction sites must operate at 110 V. Cordless drills are battery-powered.
Drill bits  Figure 5.13 Drill bits	HSS drill bits come in 1 mm to 12 mm sizes and can be used to drill wood or steel to a given diameter.
Powered nailer  Figure 5.14 Nail gun/powered nailer	Nail guns are widely used on site, particularly for new builds. There are three different types: • Pneumatic – with air hoses and portable compressors. • Cordless – with gas fuel cells, spark plugs and rechargeable batteries. • Cordless with rechargeable battery only. There are also nailing guns that will fix thin timber to hard surfaces, such as masonry, concrete or steel. Powered nailers are ballistic fixing tools and you must be certificated to use it.
Tape measure  Figure 5.15 Measures	Tape measures or rules can be lockable with return blades and belt clips. The metal versions are preferred as they are less likely to stretch. A carpenter's wooden ruler is known as a carpenter's rule. It is useful for measuring and marking smaller objects. Some are made from boxwood while others are in engineering plastic and are virtually unbreakable.
Laser level  Figure 5.16 Laser level	Laser levels have become very affordable. Originally they were expensive and only used by surveyors. They can be fixed to tripods and give very accurate levelling measurements. Another useful tool for setting out straight lines is the chalk line reel.

Tool or equipment	Use
Try square  Figure 5.17 Try square	It has a wood, plastic or metal stock and set at right angles to it is a ruled, metal straight edge blade. It is used for measuring and marking square work and for testing right angles. Similar to this is a combination square, which enables a 45° angle setting, and has an adjustable sliding blade.
Planes  Figure 5.18 Planes	All woodworking hand planes are numbered according to their width, length and or use. A No 4 (smoothing plane) is the most commonly used plane. A No 5 is a jack plane, a No 6 is a fore plane, a No 7 is a try plane and a No 8 is a jointer. There are also planes marked as ½, which denotes a plane that is wider than standard. Therefore a 4 ½ is the same length as a 4 but wider. Other planes that the carpenter/joiner should carry include a block plane, rebate plane and a carriage plane (badger) This is a small selection of the planes available.
Water level  Figure 5.19 Water level	This is a length of hose that has a transparent tube at each end. The hose is filled with water but must be clear of trapped air. It is an ideal resource for checking levels in distances of over 30 m. It is a simple device, as water will always find its own level. A great advantage is that you can take levels around corners or obstructions.
Sliding bevel  Figure 5.20 Sliding bevel	This is an adjustable gauge. The handle is made of either wood or plastic. Connected to the handle is a metal blade that is secured by a thumbscrew or a wing nut. The bevel is used to set and then to transfer angles. The blade pivots on the thumbscrew or wing nut so it can be locked at any angle required.
Spokeshave  Figure 5.21 Spokeshave	This is a very traditional tool, which has two handles and a central blade. It is used to cut down or shave timber to fit and to deal with uneven surfaces. The two handles allow the carpenter to apply differing amounts of pressure or to maintain a steady stroke with good control.
Circular saw  Figure 5.22 Circular saw	These are mainly for ripping and cross-cutting. Cordless circular saws are not usually powerful enough for constant use due to the limitations of their rechargeable batteries.

Tool or equipment	Use
Jig saw 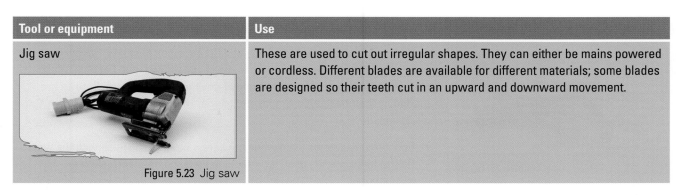 Figure 5.23 Jig saw	These are used to cut out irregular shapes. They can either be mains powered or cordless. Different blades are available for different materials; some blades are designed so their teeth cut in an upward and downward movement.

Table 5.1

Materials and fixings

Various materials are used for first fix. Many of these you will become very familiar with over time. Some are commonly used fixings while others have more specialist uses. These are detailed in the Table 5.2.

Fixings	Uses
Plugs Figure 5.24 Plugs	Plastic wall plugs are screw fixing devices. They are usually made either from nylon or polythene. They are colour-coded to match screw gauge sizes.
Nails and pins Figure 5.25 Nails	There is a wide variety of different types of nails and pins, not only for different jobs but also in different sizes and shapes: • Round head wire – for first fixing, usually 75 mm × 3.75 mm or 100 mm × 4.5 mm are used. • Lost heads – generally 50 mm × 3 mm and 65 mm × 3.35 mm and are mainly used for fixing floorboards. • Ovals – mainly used for second fix on architraves and doorstops. • Cut clasp – made from mild steel and generally used for second fixings, particularly to fix directly into mortar joints, bricks or blocks. • Cut floor brads – mainly for fixing floorboards. • Annular ring shank – one of their many uses is for fixing wind bracing to trussed rafters or chipboard flooring. • Grooved shank – a reasonably new development, lightweight and with good grip. • Panel pins – mainly used for second fix carpentry such as fixing glazing beads. • Masonry nails – ideal for fixing into sand and cement rendering, mortar joints and blocks.

Fixings	Uses
Screws  Figure 5.26 Screws	As with nails there is a wide variety of different types of screw, not only in terms of gauge and length but also in the shape of the head. Each particular type of screw can also have a variety of driving slots or recess. Screws are graded by their head type, length and gauge. Their uses are countless but could include: • Countersunk screws used when the screw needs to be flush with the work. • Raised head screws used for attaching metal components, such as ironmongery. • Round head screws used for fixing sheet material to timber. • Mirror screws have a thread in the head, which can take a decorative dome. • Pan head screws are useful for fixing sheet material.
Joist hangers  Figure 5.27 A type of joist hanger	These are U-shaped metal brackets that are used to support the ends of floor joists. They are attached with nails to a wall plate. They can also be walled or 'built in' to the brickwork. Types vary according to use; for example timber-to-timber, timber-to-masonry, etc.
Adhesives  Figure 5.28 Adhesives	Adhesives are used to bond materials together. There are two main classes: • Thermoplastic – sets when it cools and will soften if solvent is applied to the glue or it is reheated. • Thermosetting – sets and solidifies as a result of chemical reaction but this cannot be reversed. PVA adhesives bond through absorption, can be used internally and externally where specified and are a good general purpose adhesive suitable for hardwoods and softwoods, plywoods and other manufactured boards. Contact adhesives work through evaporation of solvent. This is a specific method of bonding as the adhesive is applied to both gluing surfaces, allowed to dry and then brought together so that the adhesive keys to itself. It is only used internally and is used for bonding plastic laminates. Synthetic and epoxy resins adhere through a chemical reaction when a hardener or catalyst is added to a resin in a specific quantity. They are used in laminating wooden beams, boat building and other external applications. This is just a small selection of different adhesives that you may see on site.

Materials	Uses
Timbers  Figure 5.29 Timbers	Some of the timber used during first fix will not be seen, so can be rough cut. Timber used in first fixing operations is usually but not always sawn. It is often stress-graded for uses such as floor joists and other structural work. More information about second fixing timbers can be found in Chapter 6.

Materials	Uses
Timber manufactured boards 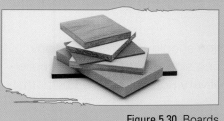 Figure 5.30 Boards	These are wood products, such as plywood, fibreboards including the widely used medium density fibreboard MDF and chipboard. Wood layers or wood fibres are glued or pressured together in manufacture to create large sheets in various thicknesses. There are also other boards used that can be waterproof or water resistant. Sterling board, or orientated strand board (OSB), is made from softwood strands that have been compressed and glued together with a resin. This is suitable for exterior work and is water resistant. It is possible to prime sterling board and give it a top coat of oil-based timber paint. It is often seen as board protection over windows in empty buildings or construction site temporary hoardings.
Plasterboard  Figure 5.31 Plasterboard	This is a panel of gypsum plaster that has been pressed between two thick sheets of paper. One side of the plasterboard is used for dry lining and the other side for finishing (such as plastering). It is usual nowadays to apply board-finished plaster to the 'white' sides of plasterboards, so they tend to be fitted white-side-out.
Insulation  Figure 5.32 Insulation	This is material that is either inserted or pumped into cavity walling or in the void of a partition wall. It is also used in roofs. It can have thermal, acoustic or fire resistant properties or a combination of all three.

Table 5.2

Selection of materials

When selecting door frames and linings at the timber yard it is essential to get 'hands on' whenever possible. Five minutes spent checking frames and linings could save hours later, as leaving the choice to the supplier's staff is not always the best option. Remember that yard staff are generally not qualified tradespeople and, although they may have a good knowledge of the stock the yard carries, they may not understand what the carpenter or joiner would be looking for when selecting materials.

Materials delivered from suppliers

It is essential that all loads are secured when they are transported to site, for example bearers should be placed between frames to prevent damage to moulds. Where ropes or straps pass across the frames, protection should be provided: corrugated cardboard is ideal. In open-back trucks tarpaulins should be used in inclement weather.

KEY TERMS

MDF

– this is an artificial board made using sawdust. MDF stands for medium density fibreboard.

Dry lining

– plasterboard is bonded to wall surfaces using the 'dot and dab' technique. Dots of plasterboard adhesive are applied to the back of the plasterboard or the wall surface and then the board is pushed onto the wall.

PRACTICAL TIP

Timber from suppliers may have been stored in ideal and controlled conditions. When it is delivered to the site it should be given a period of acclimatisation. This means allowing the timber to react to the new conditions on site – temperature, moisture and light. This will reveal any problems with the timber in advance of installation.

When taking delivery of materials and components from suppliers:

* Operatives should briefly inspect the load before unloading begins, and any obvious damage should be brought to the attention of the driver immediately.

* While unloading, a further inspection should take place and any damaged items should be put to one side to return to the supplier. This should be recorded on the delivery note and signed by both parties.

* Count the items carefully.

* If the delivery driver disagrees with your assessment do not get involved. Instead report it to a supervisor immediately.

* If you have not had the opportunity to inspect a delivery, but you are still required to sign, mark the delivery note 'Uncounted and unchecked'.

Defects in timber

Timber is a natural product so is unlikely to be perfect. Usually there are two reasons for defects or imperfections in the wood:

* it may be naturally occurring, such as a knot

* it may be caused through poor handling or seasoning of the wood after it has been cut.

The main defects and how to rectify them are outlined in Table 5.3.

Defect	Explanation and ways of rectifying
Splits Figure 5.33	Wood may naturally contain splits, or it may dry out and shrink.
Waney edge Figure 5.34	A waney edge is a defect when the wood has been cut. Some bark is left on the board and this is still on the cut plank. This may happen through the process of converting and maximising materials to avoid undue waste. To rectify a waney edge, the bark that has been left on the board would have to be removed, and cut to give a square edge on the timber.. However, waney edge is often left intentionally as a decorative feature, such as on cladding on gable ends.

Defect	Explanation and ways of rectifying
Fungal attack 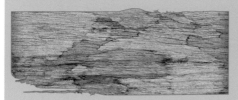 Figure 5.35	Timber that has got damp is at risk of fungal decay. For wet rot the damaged parts of the timber will need to be cut out as they will not be strong enough. For dry rot the wood needs to be sprayed with a biocide, which will kill off any strains or spores and then replaced with sound timber where necessary.
Damage caused in transit  Figure 5.36	Damaged timber materials should have been rejected at the point of delivery. It is difficult to prove that the damage has not taken place on site if they have been accepted. This means that if you are responsible for signing off a delivery, you must always check timber for damage before accepting it. Your company should have a process in place for sending back damaged goods as this may impact schedules and budgets. The exact treatment of damaged timber will depend on the defect and whether it has affected the whole delivery.
Knots 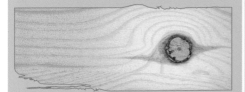 Figure 5.37	Knots are fairly common and they occur when a branch has grown out of a trunk. Knots can mean that the timber is either weakened or more difficult to work with. There are two types of knot: Dead – generally black in colour, which means that the knot is not connected to the surrounding fibres and will probably fall out as soon as the board is machined or worked with tools Live – generally brown in colour. This is where the knot is still tight and connected to surrounding fibres. If sap is bleeding from the knot it has to be sealed with a knotting solution before surface decorations can be applied. The knotting prevents resin leeching through the paint finish. An arris knot is located at a corner between the face and the edge.
Shakes  Figure 5.38	Shakes are usually found in uncut logs. There are a number of different types of shake, named after the shape in which they appear. The shakes are splits between the annual rings, or along the medullary rays, because tension has built in the tree while it was growing. If the wood is not seasoned properly and has dried out too quickly then a shake shape will appear. Shakes can also occur if the tree was allowed to fall onto a hard object when it was cut down. You cannot rectify a shake. It can be machined out at the conversion stage but is not always evident. The danger is that the timber will crack as it dries out. The wood should only ever be used where the timber is not likely to be subjected to bending.
Cupping  Figure 5.39	Cupping is a curvature across the width of a board. It is caused as a result of shrinkage occurring in relation to the growth rings.

Table 5.3

REED TIP

Everybody makes mistakes. As long as you learn from them and do not repeat them, your employer will accept this.

KEY TERMS

Profile

– this is a temporary frame.

Reveal

– a flat surface created by closing off the cavity at the opening of door and window frames.

Protecting the environment

In Chapter 3, we learned that it is important to ensure that all construction work has the least possible negative effect on the surrounding area. One of the ways of reducing its negative impacts is to keep it clean and dispose of waste in an environmentally friendly way. This means recycling as much waste as possible.

INSTALLING TIMBER FRAMES AND LININGS

There are many different types and styles of frames but there are essentially only two methods of fixing:

* Built-in frames are bedded in mortar propped level and plumb and walled in as the brickwork progresses.
* Fixed-in frames are fixed into pre-formed openings after the brickwork is complete at first fix.

The following section covers door frames, door linings, casement window frames and hatch linings.

Different types of door frames and door linings

Door frames and door linings are used for both internal and external purposes. The main difference is that the external ones are usually larger and more durable, as they have to cope with weather conditions.

The features of a door frame, regardless of whether it is internal or external, are largely the same. Door frames can either be fixed into a building as the brickwork is underway, or they can be fixed into an opening once the brickwork has been finished. It is common to use a temporary frame, known as a **profile**, if the opening is to be formed while the brickwork is underway. Once the brickwork has been finished the door frame will be permanently fixed to the brickwork.

Door linings tend to be used for internal doors. They are slightly different to door frames, as they cover the whole of the door **reveal**. These door linings are held in place with battens until the walls around them have been finished.

There are a number of different door frames and linings used in first fix operations. The following sections look at these different types in more detail.

Internal and external door casings

Interior door casings are the lining that will outline the dimension of the door. These tend to be made from softwood. Door casings can come in a variety of different widths to cope with different wall thicknesses. Internal door casings are lighter in construction than door frames. The jambs will take the weight of the door, so it is important when fixing the frame to consider the weight distribution. In other words, the jambs need to be sufficiently strong to cope with the weight and movement of the door that will be fitted into the opening.

External door casings are more robust. They also come in different sizes but are often made from hardwood. Each of the frames consists of:

* jambs – the upright components, also known as legs

* head – the top of the frame

* cill – the bottom piece of the frame (not always fitted)

* threshold – a batten fixed across a door opening that does not have a cill to prevent ingress of water and draughts

PRACTICAL TIP

All the heads of doors need to be set at the same height throughout the whole of the building. Use the site datum to achieve this. The legs of the frames or linings will extend to the floor in the case of wooden flooring but a gap needs to be left if concrete flooring has been used. Galvanised metal dowels are drilled into the bottom of the legs. These extend beyond the bottom of the leg and are concreted in when the finishing screed is applied to solid floors or beam and block flooring (or other similar construction types where there are screeds) to provide an additional fixing to the bottom of the frames' jambs or legs.

Figure 5.40 External door on frame

Figure 5.41 Internal door and lining

Fire door casings

The British Woodworking Federation (BWF) states that all fire resistant doors should be purchased from a specialist manufacturer. They recommend:

* hardwood frames with a density of more than 650 kg/m^3 for 60 minute fire doors

* softwood frames with a minimum density of 450 kg/m^3 for 30 minute fire doors.

The frames or casings should always be built in brick or block, or with a timber stud or plasterboard partition that is capable of withstanding fire up to the rate of the door. Any voids need to be filled with mineral fibre or paste.

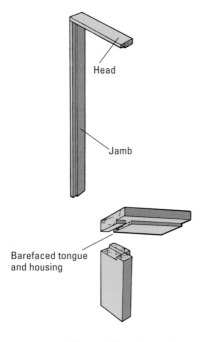

Head

Jamb

Barefaced tongue and housing

Figure 5.42 Rebated lining

Double door casings

Fixing double door casings can be slightly more complex, and even more precise calculation is necessary. There are three gaps between the doors instead of two and the jamb on each side of the opening also needs to be taken into account. The height of the opening also needs to take into account a gap for any flooring beneath the door, as well as space for the gap between the head of the lining/casing and the top of the door.

Rebated linings

Rebated linings or casings have two rebated jambs and a rebated head. The rebate is actually wider than the width of the door. This means when the door is hung it will be flush with the edges of the lining. It will have a 1.5mm clearance between the door and the stop and this will prevent it from binding.

Different types of window frames, linings and boards

Windows are designed to let in light to a building and they are also there to keep the weather out and the warmth in. There are, of course, various different types of window and each of these has particular characteristics. It is important to begin by understanding three key terms:

* Frame – this is the part of the window that supports the glass.

* Lining – this is the part of the window that fills the gap between the window and any surrounding masonry. New build windows may not have linings.

* Window board – this is at the bottom of the window and is an internal feature that caps off the top of the wall immediately below the window.

* Sash – this is a frame within the window frame that opens. It could be top hung, side hung or sliding.

Over time, a wide variety of different types of window have been developed for specific purposes. Some of them, such as dormers and skylights, tend only to feature in the roof space of a building. Other windows are more general and can be used throughout a building to contribute towards the overall style of the construction. Table 5.4 outlines some of the features of the most typical types of window.

Window type	Characteristics
Traditional	These types of window are rebated with the sash openings designed to be flush with the main frame. The joints are generally mortise and tenons.
Storm proof	These have two rebates. One of the rebates is around the main frame and the other around the casement. The main difference between these and traditional casements is that the rebates, along with other features, make them more weatherproof.
Sliding sash	There are two versions of the sliding sash window: Vertical – these have two sashes that slide up and down. They can be constructed with either boxed or solid frames. They are often referred to as being double hung sliding sash windows. Horizontal – these have windows that slide from side to side. The window is usually rectangular and the sashes slide either on a track or on hardwood runners that have been waxed.

Window type	Characteristics
Pivot	This type of window is designed so that both sides of the glass can be cleaned from the inside. In a traditional pivot window the frame and the pivoting sash are made using mortises and tenons. Storm-proof versions are also available.

Table 5.4

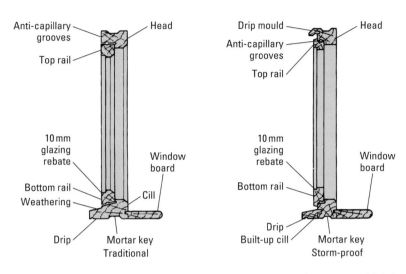

Figure 5.43 Traditional casement (left) and storm-proof casement (right)

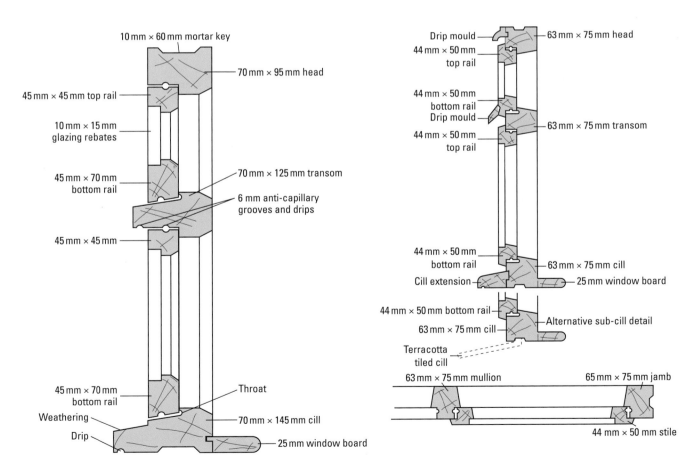

Figure 5.44 Traditional casement window (vertical section)

Figure 5.45 Storm-proof casement

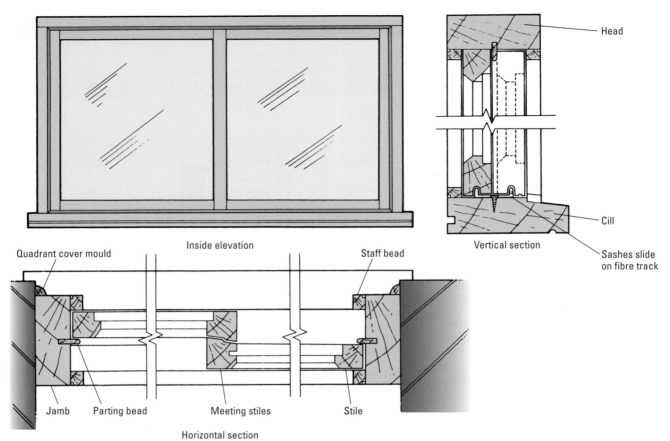

Inside elevation

Head

Cill

Vertical section

Sashes slide
on fibre track

Quadrant cover mould

Staff bead

Jamb   Parting bead   Meeting stiles   Stile

Horizontal section

Figure 5.46 Horizontal sliding sash window

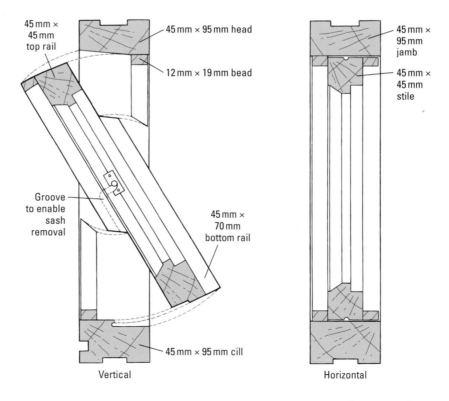

45 mm ×
45 mm
top rail

45 mm × 95 mm head

12 mm × 19 mm bead

45 mm ×
95 mm
jamb

45 mm ×
45 mm
stile

Groove
to enable
sash
removal

45 mm ×
70 mm
bottom rail

45 mm × 95 mm cill

Vertical

Horizontal

Figure 5.47 Traditional pivot window sections

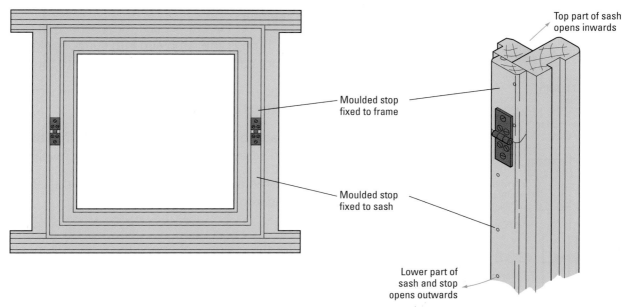

Figure 5.48 Storm-proof pivot window

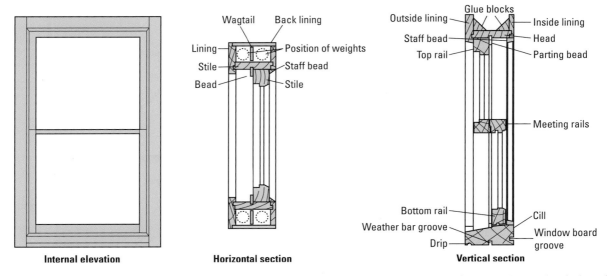

Figure 5.49 Boxed frame sliding sash window details

Methods of fixing door frames and linings in accordance with the given specification

Door frames and linings can either be built into the wall by bedding them into mortar and the surrounding wall, or they can be fixed into carefully measured openings. This happens at first fix, prior to the walls being plastered.

Fixed-in frames

Fixed-in frames are often more expensive. They are put into the openings after the bulk of the building work has been completed so that they are not damaged during that process. The lining is raised off the ground to avoid any moisture from the floor screed or any other wet products being absorbed by the wood. Any moisture that is being absorbed might make the screed dry out too quickly and this could cause cracking. Fig 5.50 shows the main ways in which the frames are fixed.

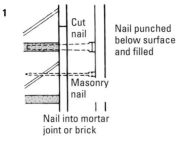

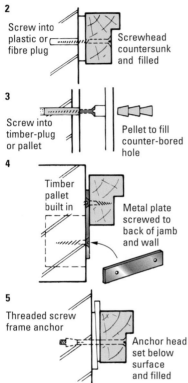

Figure 5.50 Methods of fixing frames

In the Fig 5.50 we can see that:

1 Cut nails can be driven through the jamb and into surrounding block or brick joints, or masonry nails can be driven straight into the brick.

2 and 3 Screws with plastic plugs can be driven through the jamb. Purpose made frame fixings are also available.

4 Metal plates can be attached to the jamb prior to fixing into the opening. These are then screwed into the brick or block reveal.

5 Anchors made of either metal or plastic can be fixed to the jamb and reveal and then screwed into position.

Built-in frames

This technique is used for the majority of openings. A temporary strut is put into position during wall construction. The frame can then be supported and plumbed. As the brickwork is formed around the frame, frame cramps are screwed into the jambs and then incorporated into the brickwork. The horns are cut back and treated.

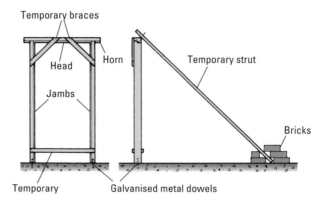

Figure 5.51 Building in a frame

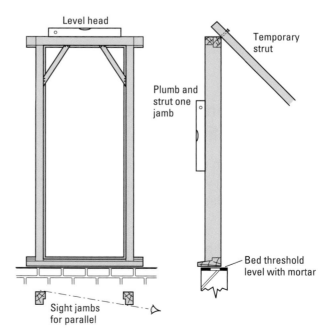

Figure 5.52 Building in a frame with threshold

PRACTICAL TIP

When using screws in softwood frames these should be countersunk. For hardwood frames the screw heads can be hidden by counter boring and pelleting (hiding the screw head with a plastic or wooden plug).

Adhesives and fixing foam

Two-part foam adhesives can also be used for the installation of frames and linings. These adhesives are often used to fix polyurethane doors. The product is mixed and activated when it is extruded using a standard silicone gun. The foam hardens quickly, it does not shrink and it expands into the gap.

It loses its stickiness in less than two minutes. It can be cut after about five minutes and becomes load bearing in half an hour. The adhesive usually hardens off between five and eight hours.

Installing timber frames and linings

There are several terms that will appear in any practical work related to installing timber frames and linings. It is important to understand these terms before you can attempt the installation of frames and linings.

Term	Internal	External
Internal or external position	If the door is inside the building, then we refer to **door linings**. An internal fire door has a frame rather than a lining.	If the door is external then it is a **door frame**.
Size	Linings for internal doors are lighter and can be up to 38 mm thick and up to 138 mm wide.	The frames for external doors are generally larger because they need to be strong and secure (50 mm to 95 mm thick).
Profile (or cross section)	Internal door linings have either planted on rebates or machined rebates.	External door frames have a rebated section with a solid stop. This makes them stronger and more weather resistant.
Proofing	Internal doors do not need to be weatherproof but they are normally draught proof.	External door frames need to have protection, particularly from rainwater. They need to be designed in such a way that rainwater will drip away from the building. They may also be fitted with a compression seal.
Construction	Door linings tend to be assembled on site.	Most external door frames will be delivered on site ready-made.
Installing	Internal linings are fit flush with the facing wall. Joints between the lining and the wall are covered by an architrave.	Most external door frames are not fitted flush with the facing brickwork. The joints between the frame and the brickwork are sealed with mastic and an architrave is sometimes put around the frame of an external door to cover the joint internally.
Wood	Linings for internal doors tend to be made from softwood.	Frames for external doors can be made from softwood but hardwood is preferable, particularly for the threshold. A hardwood cill is preferable, usually made from keruing or iroko.

Table 5.5

Frames

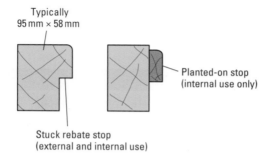

Head

Mortise and tenon

Jambs

Cill or threshold on external frames

Typically 95 mm × 58 mm

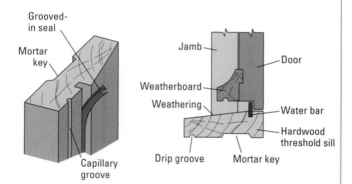

Planted-on stop (internal use only)

Stuck rebate stop (external and internal use)

Grooved-in seal

Mortar key

Jamb

Weatherboard

Weathering

Door

Water bar

Hardwood threshold sill

Drip groove

Mortar key

Capillary groove

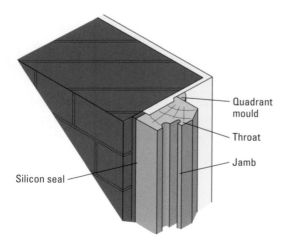

Quadrant mould

Throat

Jamb

Silicon seal

Linings

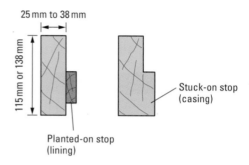

Head nailed to jambs

Housing or bare-faced tongue

Jambs

Temporary distance piece to hold jambs parallel before fixing

25 mm to 38 mm

115 mm or 138 mm

Planted-on stop (lining)

Stuck-on stop (casing)

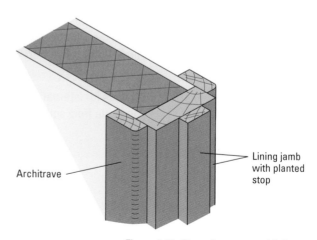

Architrave

Lining jamb with planted stop

Figure 5.53 Door frames and linings

PRACTICAL TASK

FIT DOOR AND WINDOW FRAMES AND LININGS

The following practical tasks will cover door frames, door linings, casement window frames and hatch linings.

MATERIALS

Types of fixings are determined by the background material to which the frame or lining is to be secured. This could include the following:

Pressed and round wire nails

Slotted, Phillips or Pozidriv screws and plastic plugs

Anchor bolts

Hollow wall fixings

Expanding foam

Galvanised frame cramps

Open door frames (softwood or hardwood)

Closed door frames (softwood or hardwood)

Plain linings (planted on rebate)

Rebated linings

Traditional casement window frame

Storm proof window frame

Boxed frame (sliding sashes) window frame

Pivot hung window frame

Direct glazed window frame

TOOLS AND EQUIPMENT

Hand tools	Nail punch
Panel saw	Seaming /plugging chisel
Tenon saw	Bolster chisel
Bevel edged chisels	Lump hammer
Screwdrivers	Claw hammer

Large and small spirit level, 600 and 1200

Smoothing or Jack plane

Tape measure or four fold rule

Try square and/or combination square

Carpenter's pencil

Power tools:

Cordless drill/screwdriver and assorted bits

Assorted sizes of drill bits

Countersink

Other equipment:

2 x saw horses

Scaffold boards

1. FIT FIXED-IN FRAMES

OBJECTIVE

To fit a fixed-in frame after the walls and openings are formed.

Often this will require the carpenter/joiner to construct profiles for the bricklayer. These are simple braced frames nailed together to form a template for the bricklayer to build around. This method of construction reduces damage to window and door frames as they are fitted later in the build.

PPE

Ensure you select PPE appropriate to the job and site where you are working. Refer to the PPE section in Chapter 1.

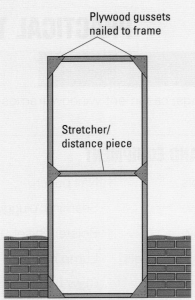

Plywood gussets nailed to frame

Stretcher/ distance piece

Figure 5.54 Example of a temporary wooden frame or profile

STEP 2 Lay the frame on two bearers and or saw horses and cut the horns, using the square, pencil and panel saw.

STEP 3 Establish reference datum point and, if possible, finished floor level (FFL).

STEP 4 Site the frame in the correct position, making sure the horizontal and vertical damp-proof course is in place. Using a long spirit level, plumb one of the jambs in both directions, sight-in the other jamb or repeat the process for the first jamb and check the head is level with a short spirit level.

STEP 1 Look at the drawing to determine the position of the frame. Consider whether the door or window opens outwards or inwards. By bearing in mind the specification and drawing the background material can be identified and the appropriate fixings selected.

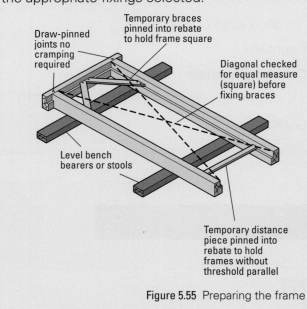

Draw-pinned joints no cramping required

Temporary braces pinned into rebate to hold frame square

Diagonal checked for equal measure (square) before fixing braces

Level bench bearers or stools

Temporary distance piece pinned into rebate to hold frames without threshold parallel

Figure 5.55 Preparing the frame

> **PRACTICAL TIP**
>
> Pack the frame against the masonry with folding wedges, pre-cut plywood packings or proprietary plastic packings. Remember to allow for plastering on faces of frames and linings that finish flush.

STEP 5 Eight fixings may be required for each jamb. These should be equally spaced using a square to keep them in line.

> **PRACTICAL TIP**
>
> A good tradesperson will hide some of the fixings behind the hinges, particularly with high class work.

> **PRACTICAL TIP**
>
> If the jambs are not parallel they are said to be 'out of wind'.

The fixings should go through the packings. If folding wedges are used, put the fixing below them, drive a nail through the frame and then both wedges will hold them in place. The two jambs should be checked for wind.

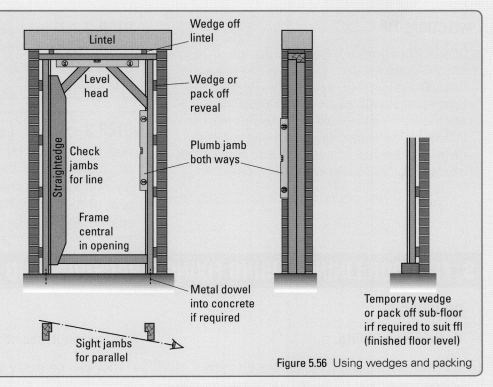

Figure 5.56 Using wedges and packing

STEP 6 After marking out, drill clearance holes using the correct sized drill bit for the gauge of screw. These should be countersunk to accommodate the head just below the surface of the frame. Using a hammer drill and masonry bit, drill through the frame into the background material.

STEP 7 Place the appropriate sized plastic plug into the hole and, using a hammer, tap it in almost flush. Place the screw into the plug and give it a slight turn to ensure it is centred then tap the screw and plug through the frame until the plug locates in the background. Tighten the screw using a hand or cordless screwdriver.

PRACTICAL TIP

Countersinking will reduce the risk of tearing up the surface with the masonry bit and will allow the screw to be extracted if necessary without breaking the surface of the timber. It also enables the effective filling over of the screwhead to allow for decoration.

2. FIT BUILT-IN FRAMES

OBJECTIVE

To fit a built-in frame before the walls and openings are formed.

This method is often used on small works where it is easier to limit damage to the frames/linings. The need for profiles is eliminated, which saves on labour and materials, but care has to be taken to ensure the frames are accurately positioned and set up, especially in the absence of finished floor levels.

STEP 1 Follow Steps 1 to 3 for *Fitting fixed-in frames* above.

STEP 2 Site the frame in the correct position and plumb the jambs in both directions, check the head for level and temporarily strut to hold in position.

STEP 3 Once the frame is held in its final position with temporary struts the frame can be built in. As the bricklaying proceeds, galvanised frame cramps are screwed to the back of the frame. These should be evenly spaced and four per jamb is the norm.

3. FIX DOOR LININGS USING FIXINGS, PACKINGS AND WEDGES

OBJECTIVE

To fix a door lining.

The techniques and tools employed to fix linings are similar to those used when installing window and door frames. However, window and door frames are normally assembled in a workshop rather than on site, while door linings are usually delivered to site in sets and have to be assembled by the carpenter/joiner.

This technique has several advantages:

* They can be easily stored on site.

* Specific linings do not have to be ordered for different sized doors.

* Ease of manual handling.

* Low cost.

STEP 1 From the drawing and door schedule determine the size of the door.

STEP 2 Measure the structural opening to ensure the lining will fit, allowing sufficient clearance to plumb and level the lining.

STEP 3 Make or assemble the lining to suit. Allow clearance when calculating the width of the lining. The lining should be glued and nailed or screwed, and the fixings should be angled in a dovetail pattern.

STEP 4 Cut battens straps to the external width of the lining. These should be fixed parallel to the head close to the bottom of the lining, front and back ensuring there is no twist in the lining jambs.

Figure 5.57 Fixing the spacing straps

STEP 5 Square the head of the lining to the jambs and temporary nail top braces.

Figure 5.58 Squaring the lining

STEP 6 Measure the width at the top of the opening, deduct the overall width of the lining and divide by two. This will give the size of the horns to be left on either side of the lining head.

STEP 7 Place the lining in opening. The two horns will centralise the lining in the opening. Now follow Steps 5 to 7 for *Fitting fixed – in frames*. Remember to allow for plastering on faces of frames and linings that finish flush.

Figure 5.59 Screwing in the lining

Figure 5.60 Fixing wedges

PRACTICAL TIP

Alternative methods of fixing linings can be adopted. The method used will be governed by both the background materials and the particulars of the construction project. For example, fixing it into timber stud partitions will adopt the same procedure as fixing linings, above; however, the lining could be nailed through packers or folding wedges into the studwork. Alternatively, if care is taken to ensure the opening is accurately constructed the use of packings can be eliminated.

Figure 5.61 The fixed lining

4. FIX DOOR LINING TO TIMBER PROPELLER WEDGES OR PLUGS

OBJECTIVE

To fix a door lining by nailing it to timber propeller wedges or plugs.

STEP 1 Assemble the lining by following Steps 1 to 4 of *Fix door linings*.

STEP 2 Rake out the brickwork seams or joints using a plugging chisel. Allocate four plugs per jamb, and if possible, choose the same level seam on both sides of the opening.

STEP 3 Cut timber plugs (propellers) and drive them into brickwork joints. Leave them long.

STEP 4 Measure the door lining and the top of the opening. Deduct the lining size from the opening and divide by two. Mark this amount on one of the top plugs (either side will do).

STEP 5 Plumb down the face of the remaining three plugs on that side and square a line across the top of each plug. This should be square to the face of the brickwork.

STEP 6 Cut the plugs and check across all four with a level to ensure they are plumb.

STEP 7 Measure the external width of the lining from each cut plug to the plug opposite, then mark as before and cut to length. Check across all four with a level as before – because they are parallel to the previous plugs they should be plumb.

STEP 8 Place the lining in the opening and check the head for level. Use packings to lift the jamb on the low side until it is level and then reduce the opposite jamb by the height of packing. Use folding wedges to pack the head tight to finished floor level. Pack off the sub-floor to finished floor level if required.

STEP 9 Fix lining to plugs using first fix brads, screws or oval nails. Remember to allow for plastering on faces of frames and linings that finish flush.

INSTALLING TIMBER FLOOR COVERINGS AND FLAT ROOF DECKING

Boards — Square-edged

Tongue & groove (T&G)

Sheets — Chipboard or OSB — Plywood

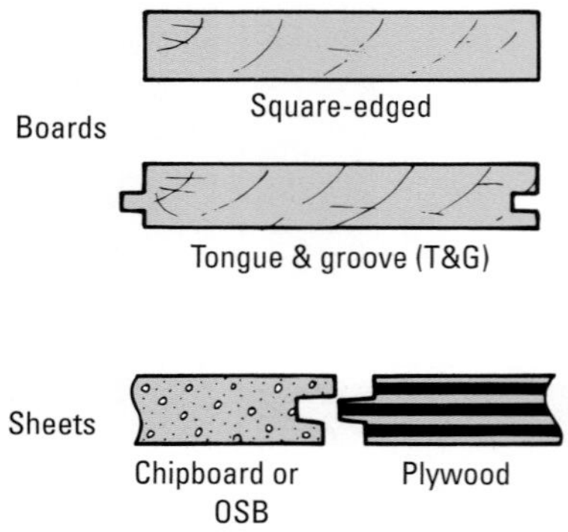

Figure 5.62 Floor coverings

Floor coverings and flat roof decking are also known as floor boarding or tongue and groove (T&G) sheet flooring. As can be seen in Fig 5.62, there are some key types of materials that are used to achieve this.

Floor coverings are normally laid:

* after any plumbing or electrical under floor work has been carried out

* after the windows have been glazed

* after the roof tiling has been completed.

The last two points are important because this means that the building is comparatively weatherproof and any floor covering will not be exposed to the weather.

Different types and sizes of timber and manufactured board joist coverings

The main materials used include the following:

* PTG (planed tongued and grooved) timber floorboards

* square edged floorboards in older buildings

* tongue and grooved flooring grade particle board

* square edged chipboard

* orientated strand board (OSB)

* flooring grade plywood.

Boards are laid at right angles to the joists. On a floor there is a 10 mm gap left from the wall to help prevent damp from being absorbed into the floorboards as a result of contact with the wall. The gap also means that the floorboards have room to expand. The gap will be hidden by the skirting.

Softwoods can be used, including red deal, whitewood and Douglas fir, in the form of tongue and groove floorboards. Chipboard and plywood are classed as manufactured types of board.

The other alternative is what has become known as engineered wood flooring, laid on top of the existing floor (not directly onto the joists). These are made from materials such as oak, walnut or maple. There are three basic types of engineered floor:

* cross-ply birch with a plywood back

* sandwich board with each part made from the same species of tree

* double layer of poplar or similar wood.

The engineered wood flooring is usually between 20 and 21 mm thick. Solid boards tend to be 18 mm thick. Engineered wood floors consist of a top layer of hardwood, which is bonded to plywood using an adhesive.

* The plywood is made up of 2 mm slices of hardwood veneers.

* The top of the board is bonded to the plywood under high pressure.

* The adhesive is then cured and the board is put into a drying chamber to reduce the moisture content.

Floor brad nailed though
face and punched in

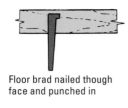

Lost heads used to secret
nail through tongue

Figure 5.63 Fixing softwood flooring

Methods of fixing joist coverings

Standard softwood flooring has a tongue and groove, which is offset from the face of the board. These boards are fixed by hammering in lost head nails or floor brads through the surface. Lost head nails are nailed through the tongue when secret nailing.

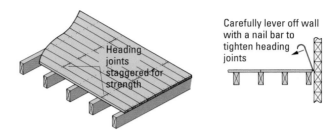

Figure 5.64 Positioning of heading joints

Heading joints are used wherever floorboards are jointed in length. The joint should always be made on a joist centre. In some cases they make use of offcuts of board and are staggered across the floor.

When boards are laid, taking into account the 10 mm gap from the outside wall, the boards are fitted, usually four to six at a time. They are pulled together using floorboard cramps. If floorboard cramps are not available, two other methods are used:

* Folding – two boards are fixed 10 mm less than the actual width of a total of five boards. The five boards are then placed into the gap and a short board is laid across them. Pressure is then applied, usually by standing on them, to press the boards into position.

* Wedging – the boards are cramped, usually between four and six boards at a time, using a fixed batten to the joist and folding wedges, as can be seen in Fig 5.66.

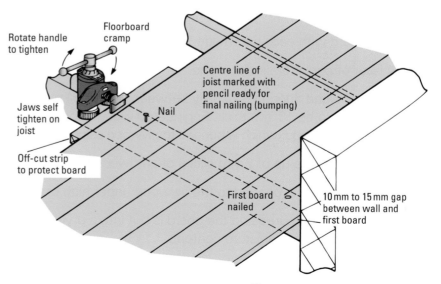

Figure 5.65 Use of a floorboard cramp

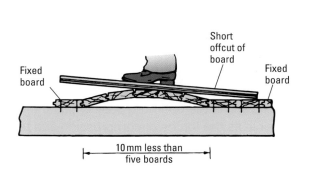

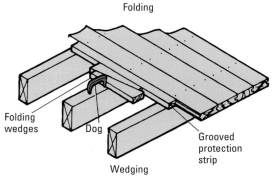

Figure 5.66 Tightening floorboards, folding and wedging

Flooring grade board is now a common material used for decking. Boards come supplied either with square edges or with tongue and grooved edges. Normally the square edged versions are laid lengthways over a joist. It is important to ensure that **noggins** are put between the joists to support the sheets.

Tongue and groove sheets have their lengths at right angles to the joists, with the short edges joining over the joists. When using tongue and grooved chipboard sheets, the shortest length permissible would be across at least two joist widths. There is no requirement for joining on a joist when 22 mm thickness is used and is glued on all edges. Where noggins are used is on the perimeter of the floor between the joist ends, and they provide a fixing for the plasterboard edges. Solid strutting prevents the joists from twisting.

It is normal practice to ensure that the joists have been specifically spaced out to match the size of the sheets. The most common spacing is 400 mm centres.

Plywood, OSB and chipboard sheets are fixed into place with nails. As with timber boards, a 10 mm gap is left along the wall for expansion and to prevent possible bridging of dampness from the external walls.

In the case of tongue and groove sheets, it is also common practice to run a line of PVA adhesive into the groove. This ensures that the floor is stiffened and there is less movement in the joints between the boards.

KEY TERMS

Noggin

– this is a short, horizontal beam timber that sits between the joists. It is used to carry sheet edges, either on the flooring when square-edged boarding is used, or the ceiling beneath the floor, and to stop the joist from twisting.

PRACTICAL TIP

Straight edge sheets of flooring grade chipboard are usually 1220 mm × 2440 mm. Tongue and groove sheets are usually 600 mm × 2440 mm.

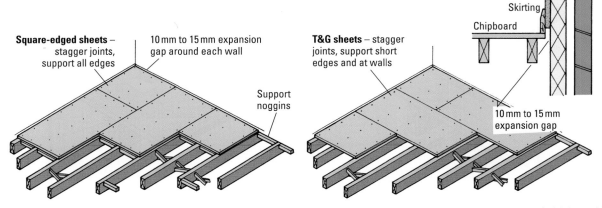

Figure 5.67 Layout of chipboard floors

Forming openings to services under floors

PRACTICAL TIP

For ease of access to services under the flooring, access traps should be fitted with recessed cups and screws.

In many cases services may have to be run under or within the flooring. This means that any cables or pipes risk having nails driven into them. This can cause short circuits or flooding if it is not carefully managed. Where possible, services should be run through the centre of joists. This is called the neutral axis and is the optimum position, as the joist is weakened as little as possible as the forces of compression and tension are equal at this point. This also allows all flooring materials to be fully fixed, although maintenance traps should be formed routinely.

Areas where there are pipes and cables should have a board section that clearly states 'no fixing'. It is also good practice to build in access points so that these cables and pipes can be reached at a later date if there is a problem.

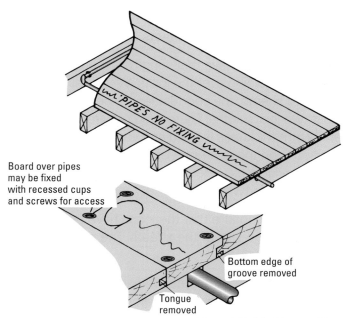

PIPES NO FIXING
Board over pipes may be fixed with recessed cups and screws for access
Bottom edge of groove removed
Tongue removed

Figure 5.68 Marking position of services

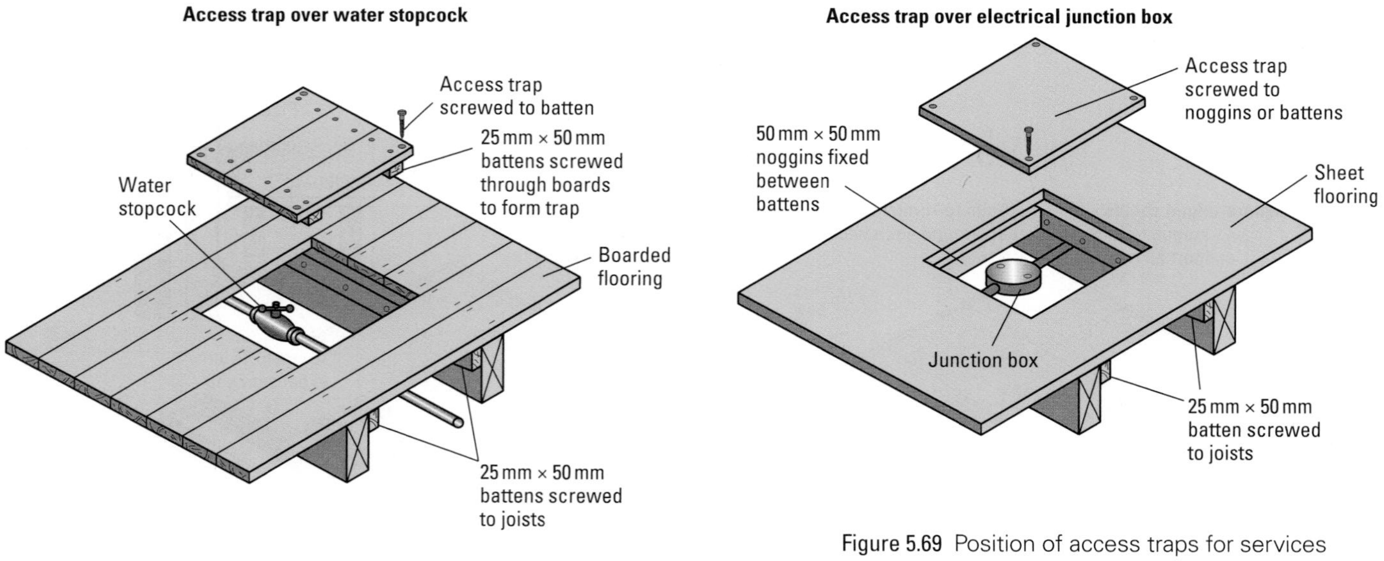

Access trap over water stopcock
Access trap screwed to batten
25 mm × 50 mm battens screwed through boards to form trap
Water stopcock
Boarded flooring
25 mm × 50 mm battens screwed to joists
Access trap over electrical junction box
Access trap screwed to noggins or battens
50 mm × 50 mm noggins fixed between battens
Sheet flooring
Junction box
25 mm × 50 mm batten screwed to joists

Figure 5.69 Position of access traps for services

Installing timber floor coverings and flat roof decking

Hardwood strip flooring is a decorative feature. It is visible in the finished building and not covered. This has also become the case with engineered woods, which are designed for decorative purposes.

Most other types of conventional flooring, particularly sheet flooring, is covered by another material for the final finish.

Materials that cover flat roof decking will be covered by a weatherproof material to insulate the property and it is likely that the decking itself will have insulation either below or above it as additional protection.

To create a functional floor, or deck, it is essential to cover the supporting joists using either manufactured sheet materials or floorboards (either softwood or hardwood). The fitting and fixing techniques for these materials are different for each and this will affect the tools required to complete the project.

Tongued and grooved particleboard sheets

These sheets are graded specifically for flooring applications and can be purchased in different thicknesses and sheet sizes. 2400 mm × 600 mm is the most commonly used, usually in chipboard. Square-edged sheets are sometimes used and these can be formed from chipboard or manufactured boards.

Timber floorboards

Timber floorboards are planed, tongued and grooved (PTG) and are available in both softwood and hardwood. Softwood is more common. Square-edged random width boards are often encountered in older buildings.

Flooring grade plywood

This can be obtained in various grades such as water boil proof (WBP) and marine, which is preferable when covering joists in an area that may be subject to high moisture, such as bathrooms and kitchens.

Flat roof decking

Softwood timber boards are rarely used in modern-day construction but, if they are to be used, they should be laid on cross battens, the length of the boards corresponding to the fall of the roof. Alternatively they can be laid diagonally to prevent water holding on the boards as they cup.

Sheet materials are more commonly used in modern construction, and bitumen impregnated boards are often used. They are laid in much the same way as for floors. They should be screwed or nailed using annular ring shank nails.

The following practical task describes how to fix and fit a tongue and groove floor.

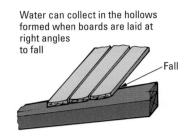

Water can collect in the hollows formed when boards are laid at right angles to fall

Fall

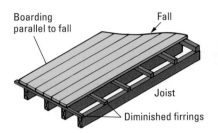

Boarding parallel to fall

Fall

Joist

Diminished firrings

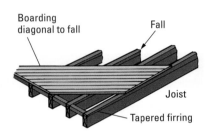

Boarding diagonal to fall

Fall

Joist

Tapered firring

Figure 5.70 Timber boarding for flat roofs

FIT AND FIX TYPES OF FLOORING

The following practical tasks will cover tongue and groove chipboard floor, softwood floorboards, and fitting access traps into both.

TOOLS AND EQUIPMENT

Cross-cut hand saw	Nail bar
Portable circular saw	Floorboard cramps
Jig saw	Claw hammer
Tape	Flat bits
Pencil	Punch

Cordless screwdriver, nail gun or manual floor nailer

PPE

Ensure you select PPE appropriate to the job and site where you are working. Refer to the PPE section in Chapter 1.

5. FIT AND FIX A TONGUE AND GROOVE CHIPBOARD FLOOR

OBJECTIVE

To learn how to fit and fix a tongue and groove chipboard floor.

STEP 1 Stack the sheets in a convenient area of the room near where they are to be fitted but away from your starting point to avoid double handling. It is good practice to do this a couple of days before work starts to allow the sheets to condition. However, in the real world this is not always possible!

STEP 2 Measure the room's width and length. (For the purposes of this example the length of the room is in line with the joists but this is not always the case.)

Divide the length of the room by the width of one sheet. This will give the number of full sheets and the remainder will be a part sheet (or ripping). If the ripping is too narrow it loses its structural stability and may break very easily. While making the calculation it is essential to allow for the 10 mm gap at both ends of the room.

PRACTICAL TIP

Example

The length of the room is 4,246 mm.

Deduct 2 × 10 mm gaps and divide this figure by the width of one sheet (600 mm). This gives a requirement for 7 sheets with a ripping of 26 mm.

Length of room = 4,246 mm

Deduct 2 × 10 mm gap = 4,226 mm

Divide by 600 = 7.043 = 7 sheets

7 sheets @ 600 = 4,200 mm

4,226 mm − 4,200 mm = 26 mm

Having calculated the ripping and found that it is too narrow, simply divide the ripping by 2 (26 mm ÷ 2 = 13) and add this to half a sheet width (300 mm) = 313. The first and last sheets will now be of equal width.

STEP 3 Divide the width of the room by the length of a sheet (2,400 mm). It will be rare that this will work out exactly; unlike the length of the room the joist spacing governs the way sheets are laid across the width. It is always good practice to have a joint on a joist; however, manufacturers now allow joints to be between joists, if all joints are glued. Assuming that our example is not to be glued, any cut board should be supported on at least two joists in its length.

Chipboard flooring should be staggered (stretcher bond). Start with a cut as close to a half sheet as possible that will leave a cut at the other side of the room that is at least the width of two 400 mm-spaced joists.

STEP 4 After you have cut the first sheet to length and width, lay it and the rest of the first row loose – do not fix at this stage. The sheets should be square to the joists; do not be tempted to push the sheets parallel to the wall.

STEP 5 There should be a gap of approximately 10 mm between the sheets and the wall along the full length of the first row. If this is not the case, for example if the wall is uneven or not square – then you will need to cut this first row of boards accordingly. Do this by setting a pair of compasses to the widest gap along the wall, then running the point of the compasses along the wall. The pencil will then mark the line to be cut along the face of the sheets. This is called scribing.

PRACTICAL TIP

A block of wood cut to the widest gap could be used as an alternative to compasses. Make the cut with a hand saw or jig saw.

STEP 6 Having made the cut, lay the sheets with a 10 mm gap, ensuring that they are square to the joists, then fix using lost head nails, annular ring shank nails or screws.

STEP 7 Start with a full sheet on the next row and continue staggering the joints on each subsequent row until you have laid the penultimate (second-last) row.

STEP 8 Measure the gap between the wall and the penultimate row. If it is consistent along its length, simply deduct 10 mm and cut the sheets to size. If the gap is not consistent then the last row of sheets will need to be scribed.

STEP 9 If scribing the last sheet, lay the sheet to be scribed parallel over the penultimate row and measure the overlap. Add the clearance gap of 10 mm to this measurement, set compasses to this figure, and temporarily nail the sheet to avoid it moving while scribing.

STEP 10 Extract any temporary nails and fix the last sheets, use a nail bar between the wall and the last sheet to ensure the joint slides home fully.

6. FIT AND FIX SOFTWOOD TONGUE AND GROOVE FLOORBOARDS

OBJECTIVE

To learn how to fit and fix a tongue and groove softwood floor.

STEP 1 Stack the sheets in a convenient area of the room near to where they are to be fitted but away from your starting point to avoid double handling. They should be stacked face up. Floorboards have a tongue and groove that is offset – this identifies the upper face.

STEP 2 Measure the width of the room and deduct the clearance from this measurement at approximately 10 mm on all walls.

STEP 3 Lay the first row square to the joists, not parallel to the wall. It may be necessary to scribe the first row as for sheet flooring.

PRACTICAL TIP

It may also be necessary to join the boards in length. This is referred to as a heading joint and should be made over a joist.

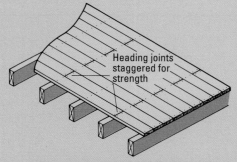

Figure 5.71 Positioning of heading joints

STEP 4 Lay a further five to six rows without fixing. These boards can then be cramped using a pair of floorboard cramps; a sacrificial board (a piece of waste wood) should be placed between the cramps and the last board. Alternatively the boards can be tightened using either of the methods shown below.

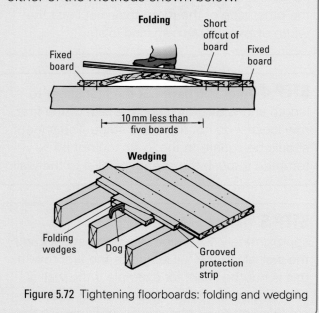

Figure 5.72 Tightening floorboards: folding and wedging

STEP 5 Nail the cramped boards using 65 mm lost head nails.

STEP 6 Continue cutting, fitting, cramping and nailing until the boarding is within one board of the wall.

STEP 7 The last board may require scribing to the wall and should have a 10 mm gap as for all other edges.

STEP 8 Using a nail bar, lever the last board off the wall to tighten the joint and fix using lost head nails.

7. FORM AN ACCESS TRAP INTO THE FLOOR

OBJECTIVE

To learn how to form access traps into floors to provide a means to install services and repair utilities, such as water, gas and electrics.

STEP 1 Determine the size and position of the trap.

STEP 2 Cut through the flooring between the joists.

STEP 3 Cut returns at 90° the first cuts to form a square or rectangular opening.

STEP 4 Screw 50 mm × 25 mm timber battens to the sides of the joists, extending past the length of the opening.

STEP 5 Fix 50 mm × 50 mm noggins between the battens; these should sit under the ends of the opening by 25 mm.

STEP 6 Screw the trap in position; ensure that the screws are below the surface and that the trap sits flush with the surrounding floor.

Figure 5.73 Dropping the trap into position

Figure 5.74 A finished trap

8. FORM AN ACCESS TRAP INTO TONGUED AND GROOVED FLOOR BOARDING

OBJECTIVE

To form an access trap into tongue and groove boarding.

STEP 1 Determine the position and size of trap.

STEP 2 Cut through the floorboards between the joists.

Figure 5.75 Cutting through the floorboards

STEP 3 Cut through the tongue of a floorboard along the length of the board between the first two cuts.

STEP 4 Lift the floorboards out of the opening.

STEP 5 Fix 50 mm × 25 mm battens on to the sides of the joists. These should extend beyond the edges of the opening under the floor at both sides.

Figure 5.76 Fixing battens on to the joists

STEP 6 Screw 50 mm × 25 mm bearers to the underside of the floorboards to join them together.

STEP 7 Screw the ready-made access panel into position, ensuring that the boards are flush with the surrounding floor.

CASE STUDY

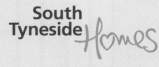

South
Tyneside *Homes*

South Tyneside Council's
Housing Company

Getting the first fix right

Michael Gaffney is a joiner in the final year of his apprenticeship at South Tyneside Homes.

'I mostly work on kitchens, going into people's houses and updating their units, worktops and floor coverings. We sometimes do adaptations for people with disabilities, for example, converting a bathroom so it has a walk-in shower. I also work on empty homes, where if a tenant were to move property or pass away, the property has to be brought back up to certain standards, a reasonable standard of living. I also get to do some bench joinery in the shop, making doors and a set of stairs recently.

I fitted my first staircase a few weeks ago; though we'd rarely get to do that because a lot of the new build work we do is bungalows. So when you get opportunity to build one, you should take it.

First fixing is the stuff I like the most because you're working from the ground up. You see things start to take shape, which is satisfying. You can more easily see that's your own work, rather than a little bit of work in the corner. You've actually built something that's substantial. First fixing progresses the job most quickly, and you see a big difference. I think it takes a little bit more skill as well because everything's got to be in the right place on the first fix. Everything that's done after that falls back onto how well the first fix has been done. Also, it's important to get it right in case it's not you who's doing the second fix, for example, when you might think you'll come back to something later at that stage. This would have a knock on effect, so it's vital to get it right and finished at the first fix.'

ERECTING TIMBER STUD PARTITIONS

A partition is a wall that divides up a large space into smaller spaces. There are two main types of partition wall: load-bearing and non-load-bearing. In modern construction, partition walling is rarely load-bearing; however, when carrying out renovation and restoration work, always investigate before removing walls in case they are load-bearing. This could be checked with a structural engineer if necessary.

There are two main types of partition wall construction:

* prefabricated
* in situ.

Timber stud partitions are the usual way in which a building can be divided up into separate areas. There are some key terms that need to be understood:

* Stud partition – this is a lightweight, non-load-bearing, timber-framed wall.

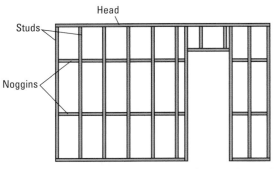

Figure 5.77 Components of a partition wall

* Stud – these are the vertical or upright members in a timber stud partition.
* Noggins – these are horizontal members that hold the studs parallel to maintain centres and to provide a fixing for the plasterboards.
* Head – this is the horizontal member that is fixed to the ceiling and forms the top of the partition.
* Sole – this is the horizontal member that is fixed to the floor and forms the bottom of the partition.

Some stud partitions are actually delivered on site ready-made. This is a perfectly workable arrangement, assuming that the spaces into which the partitions need to fit are both plumb and square. It can only be done as the storey levels are constructed and before the floors go in.

PRACTICAL TIP

Prefabricated timber stud partitions are usually made in the rooms where they are to be fixed and are slightly smaller than required to allow them to be stood up and positioned. In situ refers to a stud partition that is built piece by piece in its final position.

Types and sizes of timber used to construct partition walls

All the parts of the partition are usually made from sawn timber, which is either 100 mm × 50 mm or 75 mm × 50 mm. Most of the partitions are made on site.

Properties of materials required to cover partition walls

Table 5.6 outlines the main properties of materials that are used to cover partition walls.

Materials	Properties or uses
Framing brackets	These are also known as framing anchors. They are used to reinforce butt joints. In reality they are quite rarely used, as partition wall building specifications rarely require them.
Nails	Nails are not only used to fix the partition wall frame together, but also clout nails can be used to fix plasterboard to the partition walling frame.
Plugs	These are used when fixing the partition frame into brick or block work.
Screws	Screws could be used as an alternative to nails in fixing the partition walling together, or as a method of fixing the partition walling to brick or block work, along with the use of wall plugs.
Insulation	Should normally be fitted as standard in partition walling. The insulation may need to have additional properties, such as acoustic, fire or thermal resistance.
Horizontal cladding	The first length needs to be fixed perfectly level.
Vertical cladding	The cladding is fixed plumb at one end of the wall and pushed up tightly at the ceiling. Gaps at the bottom will be covered by the skirting. Nails are driven through into the battening of the partition wall.
Plasterboard	2,400 mm × 1,200 mm, 1,200 mm × 900 mm and 1,800 mm × 1,200 mm sheets are the most commonly used. However, 2,700 mm × 1,200 mm and 3,000 mm × 1,200 mm are also available to order. Sheet thicknesses increase as the sheets get larger. Generally either a 9.5 mm or 12.5 mm thickness of plasterboard is used.
Plywood	Plywood, or fibreboard, can be used for interior partition walling. The joints are usually covered later with mouldings. Plywood can be easily painted or plastered.

Table 5.6

Partition walls can perform other useful functions:

* The insulated material inside the partition wall can provide acoustic cushioning and cut down on noise that would otherwise easily be heard through thin partition walling.

* The insulation material can be fire resistant. This is not only valuable for dwellings but also important if partition walling is sectioning off offices from larger areas in commercial buildings.

* Thermal – in energy efficient dwellings it is important to retain the heat in selected rooms, which are being occupied. Thermal insulation in partition walling can help achieve this.

These functions are all recognised and often required by UK Building Regulations.

Methods of fixing services within partition walls

Partition walling can also hide otherwise unsightly services such as cables and pipes. These need to be supported in some way. When possible, studwork should be drilled for services rather than cutting notches into the studs; notching should only be adopted when there is no alternative. There are ways in which this weakness can be reduced, as can be seen in Fig 5.78.

Vertical studs should not be notched: all pipes and cables should be drilled through the centre of the vertical and horizontal members.

Holes in studs should be positioned away from the top and the bottom of the stud. Any additional holes into the stud should be at least three times their diameter away from any other hole.

Notches can be put in closer to the ends of the studs. But it is important to ensure that any cables or pipes are not routed through parts of the partition that will run the risk of being punctured. This means that considerable care has to be taken to make sure that any holes bored into the partition wall to hold cupboards, skirting boards or dado rails are not likely to compromise hidden cables or pipes. As a precaution, notches and holes can have a metal plate fitted over them on the wall facing.

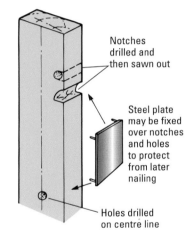

Notches drilled and then sawn out

Steel plate may be fixed over notches and holes to protect from later nailing

Holes drilled on centre line

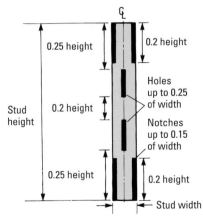

0.25 height

0.2 height

Stud height

0.2 height

Holes up to 0.25 of width

Notches up to 0.15 of width

0.25 height

0.2 height

Stud width

Figure 5.78 Hole and notched details and positioning

KEY TERMS

Framing anchor

– this is a right-angled metal bracket.

Nailed butt joint

– the end of a piece of wood is cut square and butted against the face of another. It is held in place with nails.

Skew nailing

– this is when a nail is driven into the wood at an angle to hold together two pieces of wood in the joint.

Fixing partition walls (in situ)

The timber sole plate is cut to size and placed in position. The head is then fixed to the ceiling directly above (using a plumb line). The studs can then be fitted into place. The exact distance between the studs will depend on which type of material is going to be used to cover the walls.

The noggins are added in last to hold the vertical studs parallel and provide rigidity.

The joints in a partition wall are fixed either with **framing anchors**, **nailed butt joints** or **skew nailing**.

Fixing partition walls (pre-fabricated)

Erecting timber stud partitions

As we have seen, partition walls are usually non-load-bearing and non-structural. This is very true of more modern housing. In older buildings partitions can be load-bearing.

This practical takes you through the process of forming a stud work frame. However, before work begins, a thorough survey of the location should be carried out to include:

* position of the partition: this could be from architect's drawings or direct discussion with customer

* position of any services

* position of doors and windows (both existing and proposed)

* position of any stairs

* shape of room (for example, are the walls square to one another?)

* background materials, walls, floor and ceiling

* direction of joist, floor and ceiling, if applicable.

As a result, a detailed inspection of the drawings and specification should be carried out to determine the position of the partition, the materials required and other details such as openings, coverings, trims, etc.

PRACTICAL TASK

9. ERECT TIMBER STUD PARTITIONS

OBJECTIVE

To erect a timber stud partition wall.

TOOLS AND EQUIPMENT

Hand tools:

Panel saw	Claw hammer
Tenon saw	Carpenter's pencil
Screwdrivers	Chalk line/laser
Tape measure	String line

Large and small spirit level, 600 mm and 1,200 mm

Smoothing or Jack plane

Try square and or combination square

Power tools:

Chop saw	Hammer drill
Cordless screwdriver	Nail gun

PPE

Ensure you select PPE appropriate to the job and site where you are working. Refer to the PPE section in Chapter 1.

PRE-FABRICATED METHOD

STEP 1 Mark the line of one side of the partition on the floor. Determine the square by using a large building square, the 3:4:5 method or a laser level, as many incorporate a 90° facility.

PRACTICAL TIP

Do not assume that walls are square to each other – measuring a parallel line off one wall may not give a square to another wall.

STEP 2 Take multiple measurements across the opening, both vertically and horizontally. The overall size of the frame will be the two shortest measurements, less a small allowance for clearance; this will ensure that the frame will fit the space and not become trapped by being 'long cornered'.

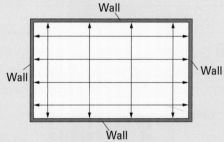

Figure 5.79 Taking multiple measurements of the room both vertically and horizontally

STEP 3 Select the straightest lengths of timber and cut the head and sole plate, remembering to allow clearance.

STEP 4 Mark out the position of the vertical studs on the head and sole plate (mark them as a pair by laying them next to one another) at 400 mm centres, allowing for any openings. 400 mm centres should be maintained above door openings and above and below window openings.

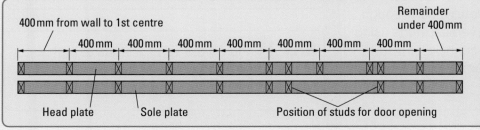

Figure 5.80 Head and sole marked out as a pair with 400 mm centre clearance maintained over door opening

STEP 5 Measure and cut the vertical studs. Remember to deduct the combined thickness of the head and sole plate from the measurement, and allow clearance.

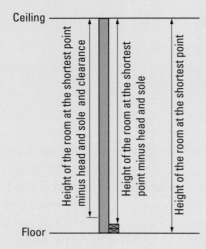

Figure 5.81 Measuring vertical studs (1)

STEP 6 Set out any horizontal members such as door heads on the vertical studs of any openings, such as for doors and windows.

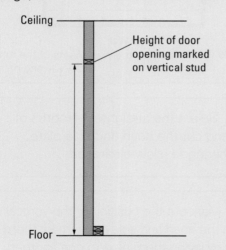

Height of door opening marked on vertical stud

Figure 5.82 Measuring vertical studs (2)

STEP 7 The stud partition should now be assembled on the floor. Dovetail a nail through the head and sole plate into vertical studs using 100 mm round-headed wire-cut nails or 90 mm first fix brads (using a nail gun). Ensure that the butt joints are flush.

STEP 8 Ensure the frame is square by checking the diagonal measurements. If necessary pull it square and attach a temporary brace across the face of the studs to make sure it stays square. Measure and cut horizontal members for doors and windows, and fix as in Step 7.

STEP 9 Mark out for noggins by measuring at 900 mm centres (assuming that the plasterboard is 1,200 mm × 900 mm) from the bottom of the frame and nail in as you did in Steps 7 and 8. It may not be possible to nail in all noggins at this stage because of the temporary brace; however the noggins can be fitted once the frame is in position and the brace has been removed. When using 2,400 mm × 1,200 mm sheets, noggins should be fixed at 1,200 mm centres to allow sheets to be fixed either horizontally or vertically and still be centred.

STEP 10 Stand the partition in position, ensuring sufficient operatives are employed for safety and to prevent the frame twisting and damaging the joints.

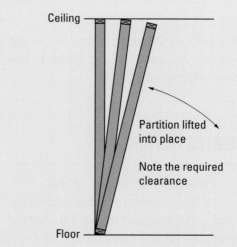

Partition lifted into place

Note the required clearance

Figure 5.83 Lifting the partition into place

STEP 11 Plumb the vertical studs at both ends of the frame and fix them to the background material.

STEP 12 The sole plate of the frame can now be pulled or pushed to the line previously marked on the floor in Step 1, and fixed to the floor at approximately 800 mm centres between the studs.

PRACTICAL TIP

An alternative to Steps 12 and 13 is to use packers and a traveller with a string line, sometimes known as string line and dollies.

STEP 13 Working from one end of the partition, plumb the studs and fix them to the ceiling as required. The partition wall is now fixed at both top and bottom.

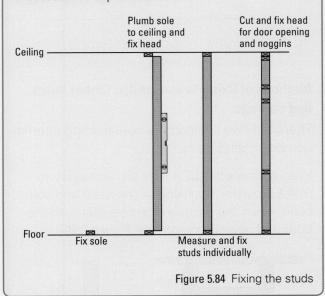

Figure 5.84 Fixing the studs

IN-SITU METHOD

STEP 1 Set out as prefabricated method by following Step 1 above.

STEP 2 Cut and fix a sole plate. Provision can be made for door openings at this stage; however, it is also perfectly acceptable to cut doorways out after the rest of the partition wall has been completed. This should be taken into account when placing fixings into the floor.

STEP 3 Mark out the sole plate at 400 mm centres and mark any door openings.

STEP 4 Measure and cut a head plate.

STEP 5 Take a short offcut and place it on top of the sole plate, then take a couple of measurements to the ceiling and cut two temporary props. These should be tight enough to hold the head plate to the ceiling when it is tapped in with a claw hammer. To avoid damaging the ceiling these should not be over-tightened.

STEP 6 Plumb a line up at both ends of the sole plate and transfer marks on to the ceiling. Connect these marks using a chalk line.

STEP 7 Lift head into position and hold it in place using temporary props. Starting from one end, fix the head plate to ceiling, pulling it straight to the chalk line while working across the length of the head. Remove temporary props.

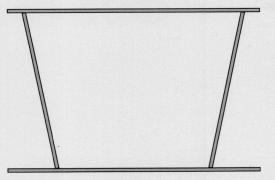

Figure 5.85 Temporary props holding head in position prior to fixing

STEP 8 Measure the distance of each vertical stud individually and proceed along the length of the partition fixing with 100 mm round headed wire cut nails or 90 mm first fix brads, stitch nailed into the head and sole plate.

STEP 10 Measure and fix noggins at 900 mm centres (assuming 1,200 mm × 900 mm plasterboard) measured from the floor up. When using 2,400 mm × 1,200 mm sheets, noggins should be fixed at 1,200 mm centres to allow sheets to be fixed either horizontally or vertically and still be on centre.

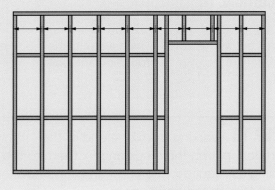

Figure 5.87 Fix all noggins

STEP 9 Measure for horizontal members such as over-door and window openings, and then fix them in position, ensuring they are level. Fix shortened studs where required to maintain 400 mm centres.

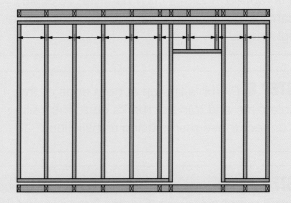

Figure 5.86 Fix all studs and then horizontal members on any openings

Methods of fixing to suspended timber floors and ceilings

There are two methods of constructing internal corners in stud walls.

With both methods, it may be necessary to make provision for fixing at the head and sole plate when the partition runs parallel with the floor and or ceiling joist (see Figure 5.88).

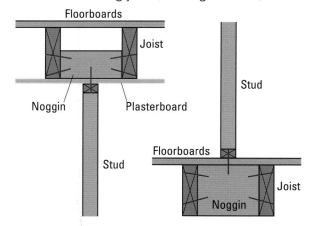

Figure 5.88 Fixing stud partitions that run parallel to joists

First erect stud partition as described using either pre-fabricated or in-situ method of construction.

METHOD 1: CORNER TRAPPING PLASTERBOARD

STEP 1 At the proposed site of the internal corner fix horizontal noggins at 400 mm centres. Alternatively fix an extra stud at the centre of the proposed corner.

STEP 2 Fix the plasterboard to the first partition.

STEP 3 Fix the last stud of partition number two through the plasterboards into noggins or stud of partition number one.

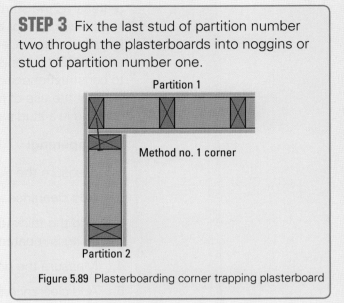

Figure 5.89 Plasterboarding corner trapping plasterboard

METHOD 2: CORNER PLASTERBOARDED AFTER STUD WORK IS COMPLETE

STEP 1 Fix partition number one, attaching a double stud where the corner is to be formed.

STEP 2 Fix partition number two to the double studs.

STEP 3 Fix the plasterboards. The double stud provides a fixing for plasterboard edges.

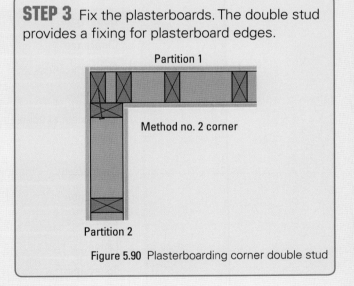

Figure 5.90 Plasterboarding corner double stud

Cladding

It is important to understand that not all studwork is clad using plasterboard. The studwork is a skeleton from which to hang the cladding, and the centres should reflect the materials that have been used. This could result in narrower or wider centres on studs, extra noggins, double head and sole pieces and so on.

Example:
9 mm boards use a maximum centre of 400 mm.
12 mm boards use a maximum centre of 600 mm.

Door and window openings

Door and window openings in stud partitions are formed from the same material as the surrounding studwork. Only the straightest timbers should be used for both vertical and horizontal members. It is common practice to allow clearance on openings; however, if planed square edge (PSE) timber is used, and care is taken, it is perfectly acceptable to construct openings to fine tolerances, so frames and linings slide in without the use of packings. The following example is for a typical door opening in a stud partition.

Door opening:

1. Measure the door width.

2. Add clearance × 2.

3. Add the thickness of two linings measured from the rebate, if the lining is rebated.

4. Measure the door height.

5. Add clearance top and bottom.

6. Add the thickness of one lining measured from rebate, if the lining is rebated.

Example:

Door width	762 mm
Clearance	6 mm allowing 2 × 3 mm
Lining thickness	36 mm allowing 2 × 18 mm
Tolerance	2 to 3 mm
Total width of opening	807 mm

Door height	1,981 mm
Clearance	3 mm top of door
Clearance	12 mm bottom of door
Lining	18 mm
Tolerance	1 to 2 mm
Total height of opening	2,016 mm

This method of construction is best adopted when the same operatives are completing both first and second fix. When adopting the more common practice of packing around openings, use calculations as above, changing the tolerance accordingly by allowing 12 to 18 mm maximum clearance. This would allow for packings between the opening and the door lining of between 6 to 9 mm on each side. Both common sizes of plywood are ideal for this.

INSTALLING STRAIGHT FLIGHTS OF STAIRS AND HANDRAILS

Most timber stairs are made off site and, in many cases, arrive on site partly assembled. On site the stairs are assembled and fixed as soon as the building is watertight.

There are various different types of staircase, but all of them have common features:

* They consist of steps, each of which has a tread and a riser.

* Each set of steps is known as a flight.

* Long flights are often broken up by landings.

* The landings are often designed so that a change of direction of the staircase can be made.

In Fig 5.91 we can see three different versions of straight flight stairs. The first shows a flight that has walls either side of it. The handrails are usually fixed directly onto the wall.

The second variation is one that is attached to a single wall and has an open side. The outer side is supported at either end by newel posts. There is a balustrade on the open side. On flights like this, with a width greater than 1 m, a handrail on the wall side is also fitted.

The final version is freestanding. It will have newel posts at the top and the bottom on each side, and a balustrade either side.

Staircases must comply with Part K of the Building Regulations.

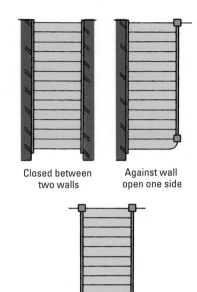

Closed between   Against wall
two walls   open one side

Free-standing
open both sides

Figure 5.91 Straight flight stairs

Different components required to form a staircase and balustrades

It is important to use the correct terms when talking about staircases. Regional variances aside, some of the terms have changed over the years.

For example, strictly speaking a staircase is the space that surrounds a flight of stairs; however, this space is now referred to as the stairwell, and staircase has become the generic term for the entire structure.

The term spindle is now commonly used but is more correctly known as a baluster with the banister (handrail) and the baluster making the collective balustrade or balustrading.

The most important thing here is that everyone involved uses the same terms. If the customer is calling balusters spindles then do the same – don't alienate the customer to prove a point!

Stair terminology

Component	Description and purpose
Baluster	Sometimes referred to as a spindle outside the trade.
	The upright piece of timber, like a small pillar or column, which fills in the gap between the bottom of the handrail and the top of the string. The gap between each baluster should not allow a sphere of 100 mm to pass through. It can be either shaped or plain.
Balustrade	The collective name for the entire assembly, which includes the handrail, newel posts, balusters, baserails and cappings.
Banister	A handrail.
Bullnose step	A quartered circle design that usually features at the bottom of a stair. It could be on one or both sides of the stair.
Capping	Capping is the process of fixing hardwood stair covers that fit or cap over the tread and riser of the stairs.
Closed string	A string that has its treads and risers housed or trenched into the inside face of the string. It is sometimes used to describe what is more accurately called the wall string.
Continuous handrail	A continuous run of handrail without interruptions, such as newel post turnings or separations. It can be straight or curved.
Curtail step	A shaped step that accommodates the volute and volute newel at the bottom of a staircase. This would be found on a stair with a continuous handrail or banister.
Cut or open string	A string that is cut to the shape of the treads and risers, to allow the tread to sit on top of the string. A return bullnose is usual on the face of the string.
Glue blocks	These are blocks of wood that are fixed to the underside of the stair. They are fixed where the riser and tread meet. The purpose of the blocks is not only to fix the risers and treads together, but to cut down on creaking and movement of the stairs.
Going	The total going is the horizontal measurement from the face of the first riser to the face of the last riser.
	The individual going is the horizontal measurement from the face of one riser to the face of the next riser. Part K of the Building Regulations states that for domestic use this should be a minimum of 220 mm.
Handrail	This is a protective component. It is designed to stop people from falling into an open stairwell. The rail forms the upper edge of the balustrade and follows the pitch or angle of the staircase.
Margin	The distance between the top string and the pitch line, which is measured at 90° to the pitch line.
Newel	Provides a means of supporting the strings, handrail and treads and risers of the stair.
Nosing	The part of the tread that extends past the individual going line. It is usually semi-circular and provides additional tread width, a finish for carpeting to curve around and a decorative finish if left uncovered.
Pitch	Also known as rake and refers to the angle at which the staircase rises to the horizontal. The pitch line is a line which connects the nosings of all the treads on a flight of stairs.
Pitch line	A notional line connecting the point where each individual going and individual riser meets in a flight of stairs.
Rake	See pitch above.
Rise	The total rise of a flight is the vertical distance between one finished floor level and another. The individual rise is the vertical measurement from top of one tread to the top of the next tread.

Component	Description and purpose
Riser	A member that forms the vertical face of the step. This could be solid timber or a composite material such as MDF. Part K of the Building Regulations states that for domestic use that this should be a maximum of 220 mm.
Staircase	The collective name for the entire structure.
Stairway/stairwell	The footprint and the space above it that accommodates the staircase.
Step	The collective name for one riser and tread.
String	This is the board where the treads and risers are fixed or cut. They can be known as wall, outer, close, cut or wreathed.
String margin	The distance, which is measured parallel to the pitch line to the top of the string.
Tread	The horizontal surface of a step. It can have a number of different shapes, such as parallel, alternating or tapered.
Wall string	The string on a staircase that sits against the wall.
Wedges	These are tapered blocks of wood. They are glued into position and used to drive the treads and risers tightly into the string of the staircase.
Winders	Steps that are used to change direction in restricted going staircases. They are narrower at one end and radiate outwards, and are often referred to as kites because of their distinctive shape.

Table 5.7

Building Regulations Part K

This part of the Building Regulations covers staircases. Part K relates to private dwellings, commercial buildings and other types of building. It also covers staircases leading to loft conversions. For these purposes we only need to concern ourselves with the domestic staircase.

Assembling staircases

There are many types of stair configuration such as straight, open riser, dog leg and geometric stairs. In order to construct a straight flight staircase the carpenter must have a good knowledge of carpentry and joinery in the first instance. The following practical exercise assumes that students will have some experience.

> **DID YOU KNOW?**
>
> The most up-to-date version of Part K of Building Regulations can be found at www.planningportal.gov.uk. There are also a number of frequently asked questions, which can give you valuable guidance when installing staircases.

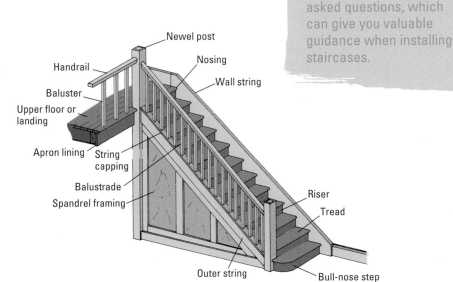

Figure 5.92 Stair construction and terminology

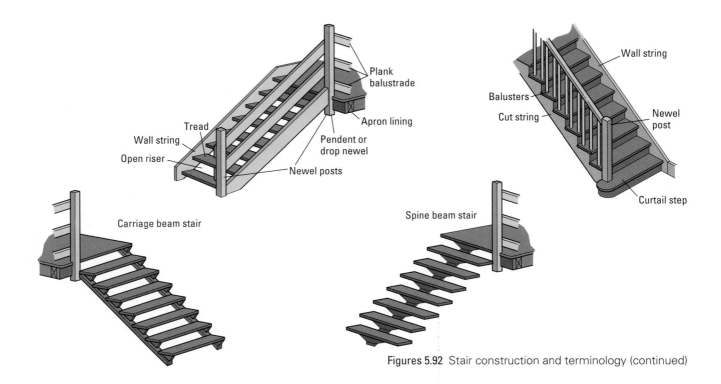

Figures 5.92 Stair construction and terminology (continued)

CASE STUDY

Coco Chanel's apartment

Below is a photo of the famous faceted mirrored spiral staircase that the fashion designer Coco Chanel (1883–1971) designed for her apartment at 31 Rue Cambon in Paris. It connected all four levels of her apartment and made it possible for her to stand in one spot and see what was happening on every floor. The staircase has a continuous string that was formed around a drum in the workshop. The balusters are arranged in clusters with large spaces between each cluster on the turns. The staircase has a very contemporary feel and allows a clearer view to each floor. However, it would be illegal under current Building Regulations in the UK.

Figure 5.93 Coco Chanel's mirrored spiral staircase

Erecting and levelling staircases

Taking site measurements

Before a staircase can be manufactured it is essential that certain crucial site measurements are taken. These measurements can be found in Part K of the Building Regulations.

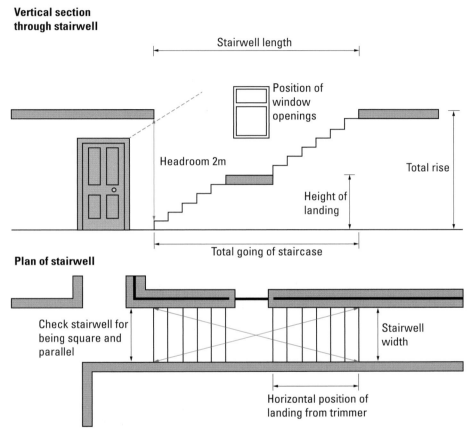

Figure 5.94 Vertical and horizontal plan of a stairwell

The maximum pitch of a domestic staircase is 42°. For a commercial application, the pitch is 38°.

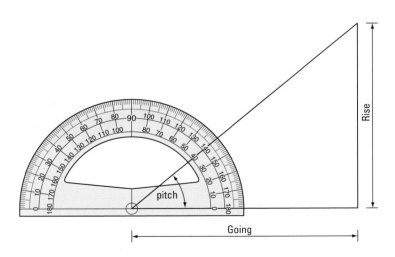

Figure 5.95 The maximum pitch of a staircase

10. ASSEMBLE A STRAIGHT FLIGHT STAIRCASE

OBJECTIVE

To assemble a straight flight staircase.

PPE

Ensure you select PPE appropriate to the job and site where you are working. Refer to the PPE section in Chapter 1.

TOOLS AND EQUIPMENT

Pencil	Wedges
Tape measure	Glue blocks
Combination square	Glue
Compass dividers	Sash cramps
Portable electric router with cutter	
Stair template	
Framing square with stair gauges	

SETTING OUT

STEP 1 Using the working drawing provided start by considering the total rise and total going. Check this with the site survey. Establish the individual riser and going dimensions and check they conform to the Building Regulations Part K.

STEP 2 Start by marking out the strings as a pair with face sides and edges

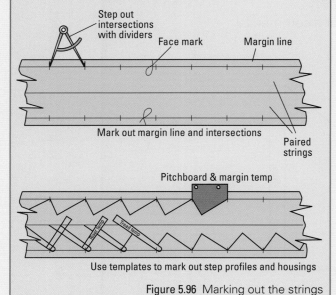

Step out intersections with dividers

Face mark

Margin line

Mark out margin line and intersections

Paired strings

Pitchboard & margin temp

Use templates to mark out step profiles and housings

Figure 5.96 Marking out the strings

STEP 3 From the drawing, establish the margin dimension and set the combination square to this. Draw this margin on the face sides using the face edge, a pencil and the combination square.

Set the framing square to the riser and going measurements using the stair gauges.

Figure 5.97 Using positioning tools

STEP 4 From the face edge, mark one riser and going dimension onto the strings. It is usual to establish the FFL (finished floor level) at the start.

Figure 5.98 Marking the risers

STEP 5 Now set a pair of dividers to the hypotenuse (longest side) of the riser and going and step this out along the string. Square these points onto the edge of the string.

STEP 6 Bring the two strings together and square the lines across and down to the margin line on the opposite string. This will keep the riser and going accurate and will avoid multiplicity of errors (creeping dimensions).

STEP 7 Use the framing square to mark out the treads and risers on both strings.

Figure 5.99 Marking out the treads and risers

STEP 8 Mark out the position of the newel post mortise and tenon joint onto the string at the top and bottom.

STEP 9 Making sure that the string is firmly secured, position the stair template. Making allowance for the router guide in front of the riser and above the going, cramp the jig in position.

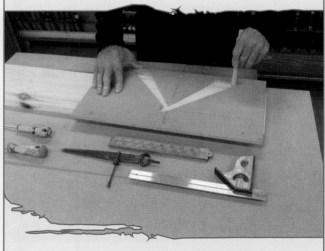

Figure 5.100 Using the stair template

STEP 10 Set the router to the correct depth and router out the tread housings on both strings.

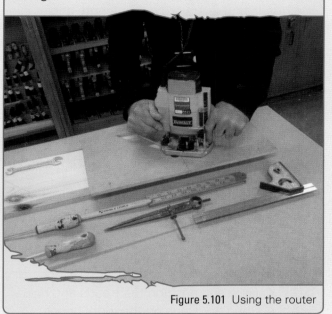

Figure 5.101 Using the router

STEP 11 Mark the mortices onto the newel posts.

STEP 12 Now mark and cut the tenons at both ends of the string.

STEP 13 Mortise the newel posts by machine or by hand.

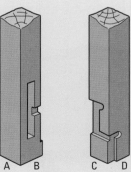

Figure 5.102 The mortise and the housing positions on the newel

ASSEMBLY

STEP 1 To assemble the staircase, make sure the treads and risers are cut to length and square, and that you have sufficient wedges, glue blocks, glue and sash cramps.

PRACTICAL TIP

Always make sure the stair is assembled dry, including the newel posts, to check the fit. Newel posts are usually fitted on site. This helps with transportation, prevents them from being damaged and allows the carpenter to make any site adjustments.

STEP 2 Glue and position the treads and risers into the string housings on the work bench.

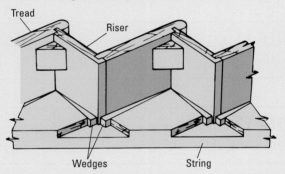

Figure 5.103 Fixing of steps onto string

STEP 3 Now cramp the strings together with sash cramps.

Figure 5.104 Cramping the strings together

STEP 4 Tap the treads into the front nosing positions and the risers into the grooves in the underside of the tread.

STEP 5 Glue and hammer wedges into position below the tread and behind the risers.

Figure 5.105 Fixing wedges under the stairs

STEP 6 Glue the triangular glue blocks in position under the tread and up to the risers.

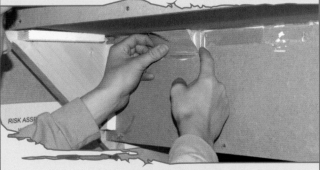

Figure 5.106 Gluing in the glue blocks

STEP 7 Check the diagonal measurement on the back of the stair. If it is square, leave the glue to set. If not, adjust the cramps to pull the stair into square and leave the glue to set.

Figure 5.107 Checking for square

PRACTICAL TIP

To work out if a staircase will be permissible, use the formula:

2 Rise + 1 Going = not less than 550 mm, not more than 700 mm.

Example calculation for a stairway for a private house

If we assume a FFL to FFL of 2.420 m and a total going of 2.880, and a figure below the maximum rise level of 220 mm, say 200 mm, we would get:

2420 ÷ 200 = 12.1 risers or 12 risers

2420 ÷12 = 201.66

If we chose a going above the minimum of 220 mm, say 240 mm, we would get:

11 goings (remember you need one less going than a riser)

Total going = 2.880 m

2.880 ÷11= 261.8

If this fits in the available plan space we can then check:

(2 × 201.66) + (1 × 261.8) = 665.12 mm

It therefore falls within 2R + 1G = 550 to 700.

Logistics

The staircase will have been fully assembled during the manufacturing process and needs to be reassembled on site.

In most situations a staircase cannot be transported to its final location fully assembled. Usually, to allow access through doorways, the balustrade and newel posts are removed, along with the top riser board, landing nosings and, if applicable, the apron linings.

Fixing considerations

Fixing a staircase is regarded as a first fix operation because it occurs prior to plastering and creates easy access to the upper floors for the workforce. It also allows the surface treatment of the wall to sufficiently reduce the amount of wood that is visible on the top edge of the wall string – this should be equal to the top edge of the abutting skirting board. If the wall string is to be fixed, it should be done at this point.

There is an argument to suggest that wall strings do not require fixing to the wall. There is sound logic to support this theory, as the string is exactly the same thickness beside the wall as it is on the open side of the stair. In addition, the stair will move and if one string can move more than the other, then the resulting forces can pull the staircase apart.

REED TIP

Good communication is vital to getting a job done correctly. Imagine a surveyor measuring up and passing on information to your team leader, who then passes it on to you – if there's any miscommunication along the line, you'll end up building the wrong thing.

Installing straight flights of stairs and handrails

You should first check the finished floor level (FFL). Stairs are often fitted prior to the FFL being finalised; if this is the case you must establish what allowance must be made for the surface covering, which could be a concrete screed, parquet or solid wood flooring.

Packing blocks will have to be used to maintain this allowance under the bottom step.

Carpets are assumed to be covering the stairs as well as the floors and no allowance need be made for them.

When installing stairs, there are several stages you need to follow.

1 Having established the FFL, the wall string must be prepared to sit on that level. Cut the string parallel to the first tread, measured down by the depth of one rise – simply measure the rise between two subsequent treads (top of tread to top of tread). Cut from the face so that any tear out is against the wall.

2 Now the plumb cut can be marked and cut. This cut forms the abutment between the wall string and skirting so it should equal the depth of skirting. Measure up from FFL and mark a 90° cut, then once again cut from the face of the string and dress it up with a block plane.

3 At the top of the wall string, there are several things to consider when marking out and cutting over the trimmer or trimming joist. The first cut is a plumb cut made in line with the back of the riser; this should meet with a square cut that is level with the underside of the flooring on the upper floor.

4 Two further cuts now have to be made: a plumb cut made from somewhere in line with the centre of the trimmer and a horizontal cut made at the top of the string the height of the skirting. This will be parallel to the upper floor. These last two cuts should be planed to ensure a good fit to the skirting abutment and a suitable finish for paint or other decoration on the horizontal cut.

5 Check the stairs for level by offering the staircase into position, packing where necessary and checking the level front to back on several treads. Ensure the stair finishes flush with the upper floor. If all is well then proceed; if not, check with your supervisor or manufacturer.

PRACTICAL TIP

Draw bore pins or dowels should be tapered at one end. The pin is driven through taper first until it has cleared the other side of the newel post.
Draw boring is a method of tightening a shoulder on a sloping tenon without the use of cramps. Offset holes between the tenon and the outside of the newel pull the shoulder in. Remember, the holes in the tenon need to be fractionally closer to the shoulder than those in the newel.

PRACTICAL TIP

The same result can be obtained by placing the newel post on the bottom string tenon, ensuring the shoulder is tight and checking for plumb with a spirit level.

6 Now the top newel can be offered up and notched around the trimmer. It can then be fitted to the upper end of the string. Glue the tenon and shoulder. The draw bore holes should already be pre-drilled using the appropriate sized dowels so apply glue and drive through the newel post.

Preparing the bottom newel post

The bottom newel post can now be fitted in the same way as the top; however, there are several things to consider.

● Is the handrail mortised and tenoned into the newels in the traditional manner? If so, it will need to be fixed simultaneously with the bottom newel.

● If the handrail is to be fixed with a proprietary fixing system then both newels can be fitted in advance of the handrail.

These operations are carried out with the staircase on its side if there is sufficient room. If space is restricted, it may be necessary to lift the staircase steeper than its finished pitch and rest it on the upper floor or landing. Struts (temporary kickers) will need to be fixed to the floor at the bottom riser to prevent the staircase slipping.

If the FFL does not yet exist, the newel may sit on the sub-floor or over-site concrete. The bottom of the newel will either need to be suitably sealed with a preservative or wrapped in building polythene sheet if it is to be screeded in.

An alternative is to cut the newel to finished floor level and to drill into the end grain. A galvanised bar or pipe can then be inserted into the hole and allowed to protrude to the sub-floor, and it can then be screeded in. This is usually done with a stronger mix than the rest of the sand and cement screed.

Fixing the newel

If the staircase sits on the FFL then, depending on the floor finish, there are numerous ways of fixing the newel. These include skew nailing or screwing, housing and skew nailing. On traditional suspended timber floors, the newel can be allowed to sit into the space between the floor joist (a trap is formed first). Noggins are fixed either side of the newel and the newel can then be bolted or screwed through. With a bit of forward planning this can be achieved before FFL and makes an incredibly strong fixing.

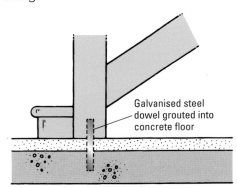

Galvanised steel
dowel grouted into
concrete floor

Figure 5.108 Fixing the newel to a solid floor

Fitting the top riser and nosing

The top riser and nosing can now be fitted into the staircase. The riser should be glued and screwed into the back of the last tread and the riser should be glued and pocket-screwed from the back into the nosing. If there is any difference in thickness between the upper FFL and the nosing then the nosing should be rebated from the underside before being fitted.

Checking the staircase

The staircase should now be dry fitted to check that everything is level square and plumb. While in this position mark the back of the bottom riser onto the floor if the stair is sitting on finished floor level. Lift the staircase back into its temporary position and screw a permanent kicker to the floor. This should be cut to fit between the string and the bottom newel.

Fixing staircases

Lift the staircase into its final position and fix the bottom newel as previously described. The top newel should be fixed into the trimmer, by pocket screwing or skew nailing using suitable length oval nails.

If a kicker is fitted the bottom riser can now be fixed through the face using screws and pellets or nails, depending on finish.

If the wall string is to be fixed, and if access to the underside is available, fix it with plugs and screws, cut clasp nails or hammer fixings, depending on wall construction.

Fixing balusters (spindles)

The design of balusters can vary greatly and there are many different types on the market. In older properties the range of designs is countless, as many would have been bespoke items.

Originally balusters would have been mortised and tenoned into the top of the string and the bottom of the handrail, but these days the handrail incorporates a wide groove in its underside that corresponds with the thickness of the balusters. A capping piece with the same groove is fixed to the top of the string. The space between the balusters is then taken up with infill pieces that are cut to the rake of the staircase.

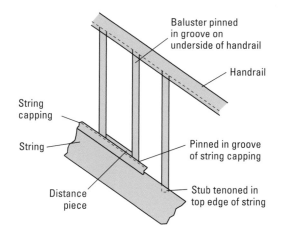
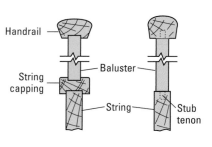
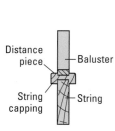

Figure 5.109 Balusters

The building regulations state that the balusters should be spaced with a gap between them that does not allow a 100 mm sphere to pass through; this sets the maximum gap at 99 mm.

The number of balusters required is dependent on the above regulation. A rough estimation can be made by taking an approximate length horizontally and dividing it up.

The balusters can be cut top and bottom at the required angles to the required length and placed in the grooves and then allowed to sit against the bottom newel. A spacer can then be temporarily fixed at the top to keep them in place. Measure the remaining distance horizontally, and divide the remaining space by one more than the number of balusters. If the gap is under 100 mm, fixing can commence; if not, add more balusters and repeat the process.

A far more accurate calculation is along the pitch/rake of the capping piece on top of the string.

PRACTICAL TIP

If the length along the capping between the newels is 3.202 m and the thickness of one baluster is 38 mm and the gap required is 99 mm, adopt the following method:

Hold a spindle in position with the required maximum spacing measured horizontally and measure the space plus the thickness of the spindle along the pitch/rake of the capping. This will give a total measurement.

Assume for the purpose of the example that this is 172 mm: 50 mm across the baluster and 122 mm across the gap.

Take the total length of the capping measured between newels and add the thickness of one baluster:

   3202 + 38 = 3230

Therefore:

   3230 ÷ 172 = 18.78

                  = 18 spaces with 0.78 mm remaining

This is not a round number so we need to try again with 170:

   3230 ÷ 170 = 19 spaces

The difference between our two attempts is 2 mm. Deduct this from our original gap (122 mm) to give us a spacer of 120 mm.

   122 − 2 = 120

If there are 19 spaces then we have one less for balusters. Therefore **18** would be required to complete the balustrade.

TEST YOURSELF

1. What is an OSBM?

   a. Ordnance Survey Benchmark

   b. Ordinary Site Benchmark

   c. Ordinary Site Bench work

   d. Ordnance Site Benchmark

2. Which of the following is an example of a collective protective measure on a building site?

   a. Hard hat

   b. Goggles

   c. Scaffold

   d. Gloves

3. Which of the following materials is commonly used to make the shaft of claw hammers?

   a. Steel

   b. Fibreglass

   c. Wood

   d. All of these

4. Which of the following types of powered nailer uses air hoses and portable compressors?

   a. Cordless with gas fuel cells

   b. Mains

   c. Pneumatic

   d. Cordless with rechargeable battery

5. What method is used to easily identify the right plug to match a screw gauge size?

   a. The shape

   b. The markings

   c. The colour

   d. The length

6. What type of adhesive sets and solidifies through chemical reaction?

   a. Thermoplastic

   b. Thermosetting

   c. Thermal

   d. None of these

7. If a piece of timber has bark left on it, what type of defect does it have?

   a. Knots

   b. Shakes

   c. Waney edge

   d. Fungal attack

8. What is the top of a door frame called?

   a. Head

   b. Post

   c. Jamb

   d. Threshold

9. In joist coverings what does PTG mean?

   a. Part Treated Groundwork

   b. Planed Tongued and Grooved

   c. Plain Timber Groups

   d. Planed Timber Groups

10. Where would you find a glue block on a staircase?

    a. On the balustrade

    b. On top of the tread

    c. On the underside of where the riser and tread meet

    d. At the edge of a tread

Chapter 6

Unit CSA–L2Occ37
CARRY OUT SECOND FIXING OPERATIONS

LEARNING OUTCOMES

LO1/2: Know how to and be able to prepare for second fixing operations

LO3/4: Know how to and be able to install service encasements and cladding

LO5/6: Know how to and be able to install wall units, floor units and fitments

LO7/8: Know how to and be able to install side hung doors and associated ironmongery

LO9/10: Know how to and be able to install timber mouldings

INTRODUCTION

The aims of this chapter are to:

* help you select resources and carry out the tasks

* show you how to erect and fix second fixing components in accordance with the given specification.

SECOND FIXING OPERATIONS

First fix operations refer to carpentry work that is carried out before the building is plastered. This includes roofing members, door frames, joists and partition walling.

The second fix carpentry tasks usually take place after the plastering has been completed. This will mean working on skirting and architraves. It has become common practice to separate the two parts of a carpenter's job. Once the first fix operations are over, other trades, such as electricians and plumbers, come in to do their work. The carpenters then come back for the second fix work.

It is at second fix stage where carpenters take on a different role, finishing off rooms and decorative features in the building. This means they could be working in bathrooms, kitchens and around the building on internal doors, architraves and skirting. They will also be finishing off the stairs by fixing handrails and balustrades. One of the other jobs will be to fix door furniture.

Additional relevant tools and equipment

At second fix some additional tools and equipment are used on site. These are outlined below.

Marking gauge
This tool is usually made from beech and has a clear, yellow plastic thumbscrew. The gauge is held with the thumb behind the pin and with the forefinger on the round surface of the stock.

Holding devices
These are cramps, such as G-cramps, sash cramps, quick release cramps (one-handed operation) that can hold pieces of timber together, or against another object. They can be used to free both of your hands so that you can fix the work more accurately.

Figure 6.1 Marking gauge

Electric mitre saw

These are often referred to as chop saws. This is because the rotating saw is brought down onto the timber. Because they speed up the operation and give consistent accuracy, many carpenters prefer to use these rather than traditional hand saws and mitre boxes.

Mitre saws are precision instruments. They can mitre cut, cross cut and cut compound bevels. They can also cut a variety of materials other than wood, including fibreglass, providing a suitable blade is fitted.

Electric router

This is another portable piece of equipment that is widely used for second fixing. Typically it will be used for:

* cutting out pockets for door hinges

* cutting out openings for letter plates

* cutting out mortices for locks and latches on doors

* jointing of laminate worktops

* cutting out the holes for handrail bolts connectors for stair handrails

* cutting apertures for sinks and hobs and cutting out slots for connecting bolts on worktops

* forming a variety of mouldings.

For some of the heavier jobs plunge routers need to be around 1300W.

Figure 6.2 Electric plunge router

Additional materials

Hanging doors, fitting kitchens and wardrobes have a priority of function over form and in actual fact skirting boards, architraves, dado rails and picture rails all have specific purposes before their aesthetics are considered. For example, dado rails are sometimes referred to as chair rails because they are fixed at a height that would prevent damage to the plasterwork or wallpaper finishes by being scraped by chair backs.

Ironmongery

This term describes door furniture. Some is decorative, like the frame or plate that goes around a letterbox, although this also protects the door aperture. Other pieces of ironmongery are to do with the function of the door, such as hinges, locks and latches. Other items are security devices and include sliding bolts and dead bolts.

Usually ironmongery is chosen to perform a particular task or to add particular detail to a door. A good example would be installing **finger plates** to an internal door.

INSTALLING SERVICE ENCASEMENTS AND CLADDING

To cover unsightly services such as electrics, water and gas and to provide an easily decorated surface, pipework needs to be encased or boxed with the overall size kept to a minimum. Access may be necessary to the pipe or cable work to provide for maintenance of the service. For example, plumbers may need to access stopcocks, pipework and valves, or electrical connections might have to be tested and repaired.

Methods of encasing services

Encasing services should be tackled in a careful way. You do not want the casing to cause unnecessary problems, particularly with the final finish of a room. The following points should be taken into consideration:

* Always try to use standard measurements of timber – if this means making the casing slightly larger than is necessary you should do so. This will make all other measurement and preparation of the room easier.

* Casing in bathrooms and kitchens that are going to be tiled should ideally be in multiples of a whole tile or half a tile; this gives a more pleasing result and makes tiling easier.

* Remember that access points may be needed. Whenever there is a junction or valve the casing needs to have a removable face board.

* Make sure you use the correct materials – for example, in kitchens and bathrooms **MRMDF**, marine plywoods and external MDF boards such as medite are commonly used.

KEY TERMS

MRMDF

– moisture-resistant medium density fibreboard.

Types and sizes of cladding

There are various different types and sizes of cladding. These are outlined in Table 6.1:

Type of cladding	Size and description
Solid panel	This is made from slow-grown, softwood timber, such as spruce. It is used for a variety of internal cladding jobs. It is V-jointed and can be installed horizontally, vertically or diagonally. It has a finished size of 8 mm × 94 mm with a coverage of 88 mm. Sizes available are usually up to 3 m.
Manufactured boards	These are either PVC or polypropylene sheet systems. Each of the sheets is joined to the next using a joining strip and an edge piece. The joints and edges are screwed to the wall. The sheets are 3 mm deep, up to 3 m long and 1.5 m wide. These include MDF, plywood, plasterboard and hardboard.
Tongue and groove boarding	This has a finished size of around 8 mm × 94 mm with a coverage of 88 mm per board. It can be used as either vertical or horizontal cladding. Usually it is available in lengths of 1.8 m, 2.4 m and 3 m.
Horizontal and vertical boarding	This type of boarding is softwood and V-jointed. It has a finished size of around 14 mm × 94 mm with a coverage of 87 mm per board. It is usually available in 1.8 m and 2.4 m lengths.

Table 6.1

Insulation

All materials have an insulation value measured in terms of the amount of thermal loss through the material. This is known as a U-value rating. The lower the U-value the better the thermal insulation properties. The insulation of properties should be carefully planned as insulation placed in the wrong sequence can cause internal problems such as condensation.

Internal walls can also benefit from insulation. Rather than make standard partition stud walling, stud wall insulation can be used. This uses rigid insulation boards. These can be fitted over existing walling. One widely used insulation material is made from a foamed plastic that is between 60 and 100 mm thick.

In the case of stud walls a metal or wooden framework is built and then filled with mineral wool fibre. The mineral wool insulation is not as effective as rigid insulation boards and this means that the filling has to be thicker.

To provide additional insulation a standard stud wall framework is built and rather than covering the framework with plasterboard it is covered with rigid insulation boards.

Soundproofing

Soundproofing, or sound insulation, is designed to cut down on the impact of sound from one part of the building to another. This is also known as acoustic separation.

Different materials have different sound insulation properties.

Soundproofing panels can be fitted on walls, ceilings, onto roof panels (to cut out rain noise) and they can also be fixed to the outside of a building. They are usually around 50 to 60 mm thick. A series of steel studs hold them in place and it is usual to have a second protective layer, or architectural wall panel, fitted over them.

Fixing timber grounds plumb and level

All panelling, whether it is a small box that encases pipes or cables or a larger piece of panelling, must be straight level horizontally and plumb vertically. The walls onto which the panelling will be fixed have to be battened. A ground is a timber batten that is fixed to masonry walls to provide a fixing. They should be fixed either vertically or horizontally depending on their application but are fixed to the structure before panelling.

As you can see in Fig 6.3 there are three types of grounds: framed grounds, separate grounds and counter-battening.

Grounds provide a fixing for surface materials that disguise very uneven wall surfaces. Using a level and a straight edge it is possible to identify surface irregularities. The grounds can then be packed out to meet the line. This will ensure the wall is plumb and in line.

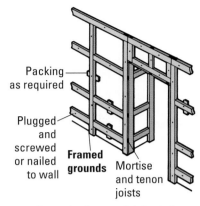

Packing as required

Plugged and screwed or nailed to wall

Framed grounds

Mortise and tenon joists

Framed using pre-fabricated mortise and tenon frame

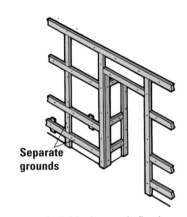

Separate grounds

Individual grounds fixed on centres

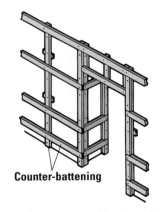

Counter-battening

Corner battered horizontally

Figure 6.3 Types of ground

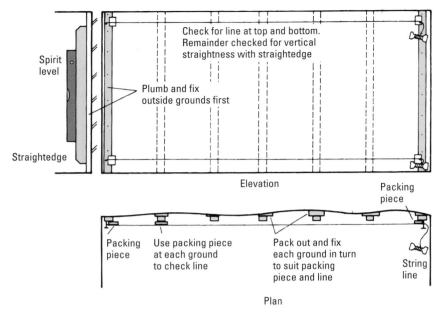

Figure 6.4 Plumbing and lining grounds

Framed grounds tend to be made off site, and are generally packed off the surface in the same way as the separate timber grounds noted above. The process of plumbing and lining grounds can be seen in Fig 6.4.

Methods of fixing cladding (internal panelling)

When you fix panels to grounds the actual fixings sometimes have to be concealed. There are various ways of achieving this, as can be seen in Fig 6.5.

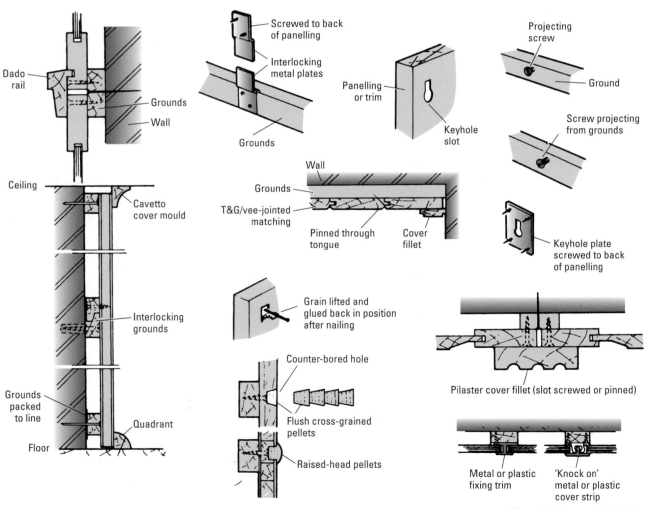

Figure 6.5 Concealed fixings

There are various ways of fixing cladding and these can be concealed or surface fixed. These are listed and described in Table 6.2.

Fixing	Use and fitting
Interlocking grounds	This technique uses either splayed or rebated grounds. These are fixed to the back of the panelling and to the wall. The panel is lowered and then hooked into position.
Slot screwing	These are keyhole shaped slots that are on the back of the panelling. The slots are drilled and prepared so that they match the countersunk head screws that have been driven into the grounds. The panel is manoeuvred into position so that the slots are over the head of the screws. The panel is then tapped into position.
Slotted and interlocking metal plates	Keyhole slotted metal plates are fixed into recesses on the back of the panels. In every other respect they work on the same basis as interlocking grounds (in the case of interlocking metal plates) and slot screws (in the case of slotted metal plates).
Pellets	Holes are bored into the panelling and then pellets or plugs are inserted into the holes and glued into place. The pellets, which are cross grained, are virtually invisible once they have been installed.
Cover fillets	The panels are simply screwed directly into the grounds. Where the surface screws are still visible, a fillet, or moulded feature, is then pinned to cover the screws. In practice these could be cornices, skirting, dado rails or coving.
Nails	This involves driving a nail through the panelling and into the ground. The nail head can be hidden using hard wax or exterior filler. In some cases a tiny piece of the grain can be lifted with a chisel or grain lifter. The nail is then driven in. The grain is then replaced and glued to cover the nail head.
Trim	These are raised and very obvious, so they are features of the panelling. They can be made of either metal or plastic.

Table 6.2

Figure 6.6 shows how internal and external angles can be handled.

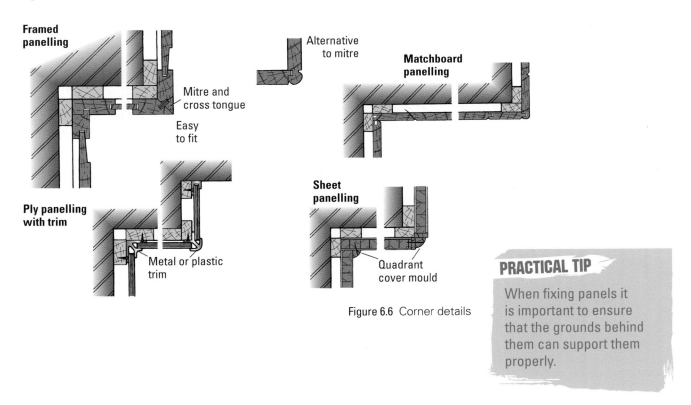

Figure 6.6 Corner details

PRACTICAL TIP

When fixing panels it is important to ensure that the grounds behind them can support them properly.

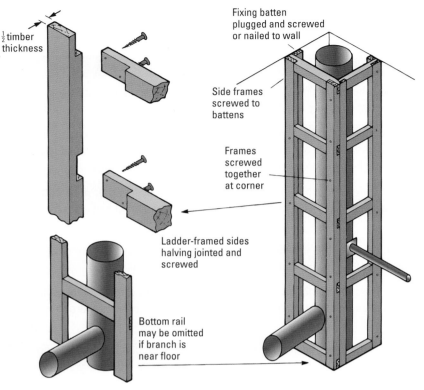

½ timber thickness

Fixing batten plugged and screwed or nailed to wall

Side frames screwed to battens

Frames screwed together at corner

Ladder-framed sides halving jointed and screwed

Bottom rail may be omitted if branch is near floor

Figure 6.7 Use of ladder frame for pipe casing

Pipe casing

For vertical pipes one of the common casing techniques is to use ladder frames. Battens are fixed to the wall and then softwood is fixed to the battens to provide a solid ladder-framed box. This can be seen in Fig 6.7.

There are issues when constructing L-shaped, U-shaped casings or where the wall surface is uneven.

L-shaped casings are used to cover services in internal corners. Begin by using a spirit level and a straight edge to mark plumb lines on the wall. The battens then have to be fixed to the wall. The sides of the case and finally the front can be fixed. It is also possible to use a ladder frame. A notch should be created where a pipe branches off. If the pipe is small then it is possible to simply drill a hole and saw the side. But for larger pipes the technique to cover this is usually by scribing and having a two-part split face, as can be seen in the diagram.

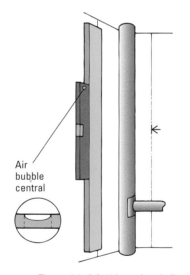

Air bubble central

Figure 6.8 Marking plumb line on wall

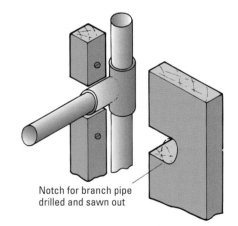

Notch for branch pipe drilled and sawn out

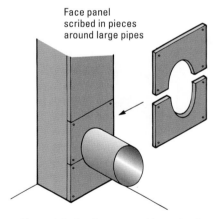

Face panel scribed in pieces around large pipes

Figure 6.9 Cutting around branch pipes

U-shaped casings are used to cover services in the middle of a wall. The plumb lines are again marked before fixing battens using the same procedures as for L-shaped casings.

For uneven walls scribing is usually required. Put the side piece of external cladding against the wall and then mark down the piece of timber using either a pencil for a small scribe or use a compass set to the maximum space identified to mark a line. You will then need to plane or cut the cladding to the line. This can be seen in Fig 6.10.

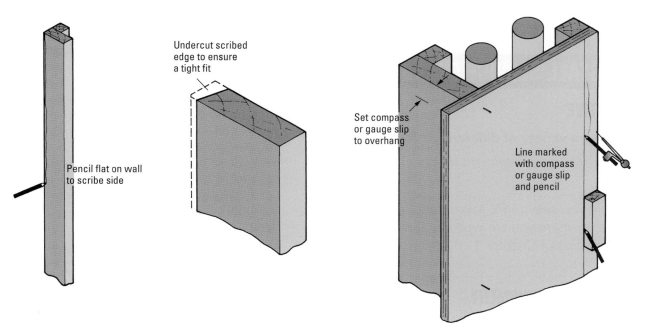

Figure 6.10 Scribing and cutting to fit an uneven wall

Covering horizontal pipes can be achieved in exactly the same way as for vertical pipes. In some cases it may be desirable to cover the facing with skirting board or to fix an overhang on the top. This would effectively provide a shelf. The three alternatives for horizontal pipe casings can be seen in Fig 6.11.

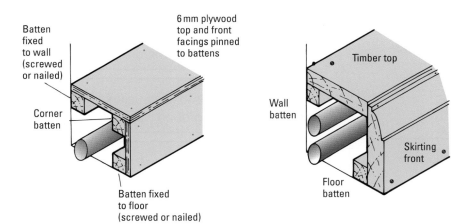

Figure 6.11 Horizontal pipe casings

PRACTICAL TASK

1. INSTALL SERVICE ENCASEMENT AND CLADDING

OBJECTIVE

To install a variety of different casings.

PPE

Ensure you select PPE appropriate to the job and site where you are working. Refer to the PPE section of Chapter 1.

TOOLS AND EQUIPMENT

Hand tools:

Tape measure	Claw hammer
Pencil	Screwdrivers
Spirit level	Drill bits
Straight edge	Countersink
Handsaw, panel and or tenon	

Power tools:

Cordless screwdriver	
Drill bits, for wood and masonry	
Chop saw	Hammer drill
Jig saw	110V transformer

PRACTICAL TIP

When using thin sheet material it may be necessary to fix front fillets to the sides, to allow a fixing for the front panel.

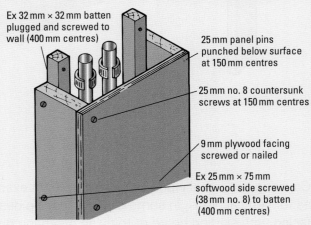

Ex 32 mm × 32 mm batten plugged and screwed to wall (400 mm centres)

25 mm panel pins punched below surface at 150 mm centres

25 mm no. 8 countersunk screws at 150 mm centres

9 mm plywood facing screwed or nailed

Ex 25 mm × 75 mm softwood side screwed (38 mm no. 8) to batten (400 mm centres)

Figure 6.12 U-shaped pipe casings

L-SHAPED CASINGS

STEP 1 Mark plumb lines on the walls using a spirit level and a straight edge, ensuring there is room for the chuck of the drill.

STEP 2 Fix battens to lines using appropriate fixings for the background material.

STEP 3 Fix the side panel to the batten.

STEP 4 Fix the front panel to the side panel and wall batten.

U-SHAPED CASINGS

STEP 1 Mark a plumb line either side of the pipework, allowing enough room to accommodate the chuck of the drill during Step 2. This is to avoid damaging the pipework or services.

STEP 2 Fix battens to the lines marked on the walls.

STEP 3 Fix the side panels to the battens.

STEP 4 Fix the front panel to the sides.

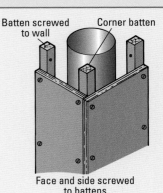

Batten screwed to wall

Corner batten

Face and side screwed to battens

Figure 6.13 L-shaped pipe casings

LADDER FRAMES

Ladder frames are often constructed off or on site to carry the finishing material for the encasement of services. They consist of simple lightweight frames, which, once fixed and clad, become extremely strong.

STEP 1 Measure up for the frames. For L-shaped casings two frames will be required. For U-shaped casings three frames will be required.

STEP 2 Cut the timber to the required lengths and widths and half lap or mortice and tenon the frames together (the latter would usually be manufactured in a workshop).

STEP 3 Screw the frames together on the front corners.

STEP 4 Mark plumb lines in the corners and fix using appropriate fixings.

STEP 5 Clad with finishing material.

SCRIBING COVER PANEL TO UNEVEN SURFACES

In many cases the sides and faces of casings will need to be scribed to fit uneven wall or floor surfaces.

STEP 1 Position the panel to be scribed against the framework or in its position. Ensure it is plumb or level depending whether it is a horizontal or vertical pipe box, and temporarily nail it in place.

STEP 2 Set a pair of compasses to the overhang and mark a line down the face of the panel.

STEP 3 Cut the panel, slightly undercutting to ensure a tight fit. It may be necessary to dress in with a block plane to obtain the required finish.

STEP 4 Fix the panel.

STEP 5 Repeat for other faces as required.

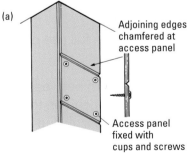

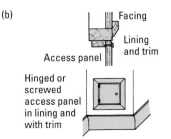

(a) Adjoining edges chamfered at access panel

Access panel fixed with cups and screws

(b) Facing

Lining and trim

Access panel

Hinged or screwed access panel in lining and with trim

Figure 6.14 Access panels

Access panels

We already know that access panels may be necessary to allow other trades such as electricians or plumbers to access cables and pipes. Usually this is achieved by creating a removable panel that is fixed into position with brass cups and screws. Sometimes a hinged door is more suitable.

Access panels often have chamfered edges. The reason for this is that the edges of the access panels are often prone to damage each time they are removed or fixed back into position. It also means that when the panelling has been painted the joints are easy to cut open with a knife, as they will have only a thin film of paint.

The framework is constructed in the normal way but the panels will have to be joined where access is required.

Options for access panels can be seen in Fig 6.14.

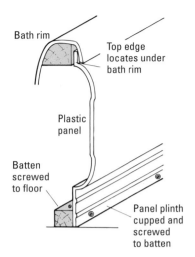

Figure 6.15 Fixing standard bath panel

Bath panels

Some baths will come as kits with standard plastic bath panels. All that is necessary is to position a floor batten. The bottom of the bath panel is fixed to the batten after the top has been located into position under the bath rim.

In other cases it is necessary to create a batten framework for the panel. Usually this is made from 25 mm × 50 mm softwood. It is usually planed all round, halved and then screwed into position.

Regardless of the type of panel material chosen, the sheet material needs to be moisture resistant. In many cases hardboard that has been faced with melamine or MRMDF can be used. In other cases match boarding is used.

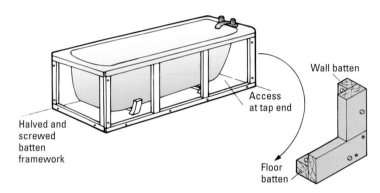

Wall batten

Access at tap end

Halved and screwed batten framework

Floor batten

Figure 6.16 Batten framework for a bath panel

INSTALLING WALL UNITS, FLOOR UNITS AND FITMENTS

In most dwellings, kitchens are laid out to a plan to make full use of the available space.

The usual rule of thumb is that there is a triangular pattern between the sink, cooker and fridge. The distance and spacing will depend on the client's preferences. The cooker is not usually next to the sink due to the fact that the cooker will usually require an electricity supply and the sink will need water. Several other factors also need to be considered:

* The need for a food preparation area between the sink and the cooker.

* There needs to be a work surface over or near the fridge.

* The cooker should not be too far from the sink.

* The need to make good use of natural light. It is common for the sink to be underneath the window.

Many units come in standard sizes. These can be bought off the shelf from a variety of suppliers. Kitchens may also have purpose-built units, and in most kitchens each work surface or worktop will have to be specifically cut to fit. This will involve taking into account the layout of the room.

Detecting and protecting electric cables and gas and water pipes

It is a sensible and necessary precaution to disconnect any electrical wiring. It is also a good idea to cap off hot and cold water supplies. The gas supply should be turned off and disconnected. When this is required you must use qualified and competent relevant trades.

Electricity
A kitchen will have a ring main or ring circuit. This will provide a number of double socket outlets above the worktops. There will also be electric sockets to provide power for washing machines, fridges and dishwashers. These are sometimes wired directly into a fused box – if so, a qualified electrician should always be consulted. A separate circuit may be necessary if the cooker or hob is electric, depending on the load of the appliance.

All electrical work needs to be carried out by a competent electrician. However a non-specialist can create the channels to conceal the cables providing no connections are made.

Tracing hidden electrical wiring can be tricky. Voltage detectors are only helpful once you have found the wire. They will tell you whether the wire is live or not. The best practice is to use all-in-one wire tracers to detect wires behind walls but extreme care must be taken when fixing units.

Many professionals will use multiwall scanners and detectors. These are able to find cables, pipes, girders, frames and even small metal objects like nails and screws.

Figure 6.17 Using a multiwall scanner

Gas
Pipes can be detected in a similar way to electrical cables. Once their routes have been identified they should be marked 'Pipes No Fixing' to ensure that no screws or nails are driven into these areas. It is highly likely that gas pipes will be concealed within stud partitions.

Water
Water pipes are another example of hidden or buried services. Pipes that are buried in walls can be difficult to find but the position of sockets, gas outlets and taps offer clues to the possible direction of hidden services. Alternatively, water services can be detected using a scanner. These should also be marked to prevent nails or screws being driven into the danger areas.

Methods of fixing wall and floor (base) units

The ways of assembling and installing standard flat pack units will vary. You should always read the supplier's instructions.

Usually there are three stages involved:

1. Assemble the units if they are not pre-assembled. Make sure that you check the contents as you unpack them. Only open one pack at a time so that you can identify which parts belong to each unit and whether there are any missing parts.

2. Carry out the installation starting from the highest point within the room. Usually you will start with a corner base unit and work outwards. After the base units are installed a corner wall unit is fitted and then you work outwards from that.

3. Once the units are in place they need to be finished. Worktops should be installed first, then cornice and lighting pelmets and finally doors and drawers, as there is a chance of damaging the doors or drawers if they are installed too early in the process. This should always be one of the last tasks, leaving only the fitting of the handles.

Floor (base) units

Each base unit needs to be level and plumb. Packing or, more commonly, legs can be adjusted to allow for any uneven floors. Each of the units needs to be flush with each other. This means making sure the top edges and the fronts are in line. Each unit is then secured to its neighbour with screws or purpose made joining bolts supplied with the kitchen.

Once this has been done the base units can then be fixed to the walls. Brackets are usually provided. You then drill and plug before screwing the units to the walls.

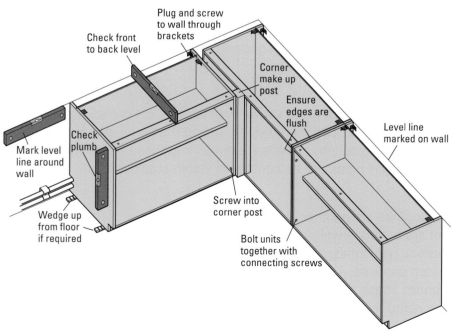

Figure 6.18 Fixing base units

Wall units

Wall units can be a mix of tall units, which are basically large cupboards, or standard wall units. In most kitchens the tops of all these wall units are level. In order to achieve this, a horizontal datum line is drawn onto the wall to mark the top of each of the units. A second line is then drawn to mark the depth of the unit below this line.

You now have the upper side and underside fixing positions marked.

Fig 6.19 shows how wall units are fixed. It is necessary to mark the wall through the fixing holes provided. These holes will then have to be drilled and plugged before the unit can be screwed to the wall. You can use the hanging brackets provided or you might need to fix packing behind the wall unit if the surface of the wall is uneven.

Once the wall units are flush they can be secured together with either screws, plugs or the connecting bolts provided with the kitchen.

> **PRACTICAL TIP**
>
> It is usual practice to have a gap of around 450 mm from the underside of a wall unit to the top of a work surface.

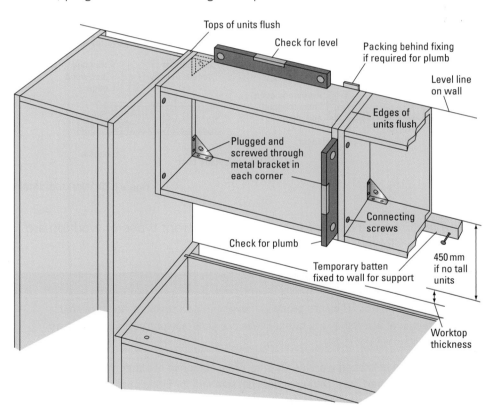

Figure 6.19 Fixing wall units

Jointing post-formed and timber worktops

Just as the styles of base units and wall units have changed over time, so too have preferences about worktops. Not all worktops are made from timber. Metal, marble, stone and other solid sheet material may need equipment and cutters that are not part of the standard kit of a carpenter, so specialist fitters may be required.

There are various different types of timber worktops, which can be seen in Fig 6.20.

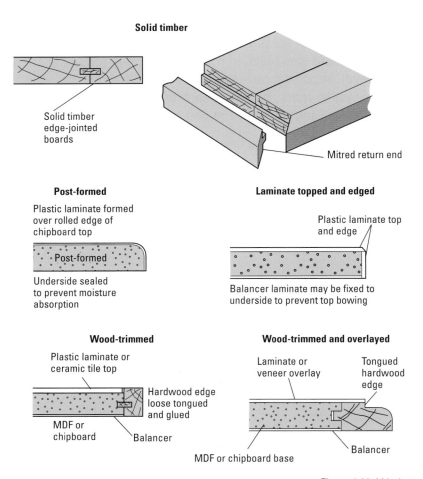

Solid timber

Solid timber edge-jointed boards

Mitred return end

Post-formed

Plastic laminate formed over rolled edge of chipboard top

Post-formed

Underside sealed to prevent moisture absorption

Laminate topped and edged

Plastic laminate top and edge

Balancer laminate may be fixed to underside to prevent top bowing

Wood-trimmed

Plastic laminate or ceramic tile top

Hardwood edge loose tongued and glued

MDF or chipboard

Balancer

Wood-trimmed and overlayed

Laminate or veneer overlay

Tongued hardwood edge

MDF or chipboard base

Balancer

Figure 6.20 Worktop details

The features and uses of each of these different types of worktop are outlined in Table 6.3.

Worktop type	Features and uses
Solid timber	These are made from jointed boards. They are prone to movement as a result of moisture so this needs to be taken into account when they are fitted. These are generally untreated and must be sealed after fitting.
Post-formed	This is the most common type of kitchen worktop surface. They comprise of either MDF or chipboard covered in a plastic laminate. Apart from cutting at the wall side and for fitting appliances such as hobs and sinks they are essentially ready for installation.
Laminated in situ worktops	These are also either chipboard or MDF and they can be covered with plastic laminate, wood veneer or melamine. Once these have been fitted into place a matching sheet material is then applied to the edging. This can then be sealed, as can the underside, with varnish or adhesive. Note that edged sheet is rarely used in modern kitchen fitting as it is labour intensive, does not give as good a finish when complete and is overly dependent on the skills of the carpenter/joiner to achieve the quality demanded.
Wood trimmed	This is another chipboard and MDF material. It also has a wood veneer, plastic laminate or melamine covering. A hardwood edging is fixed using glue and screws or it can be rebated.
Wood trimmed and overlaid	This has either a chipboard or MDF base. It has a hardwood edge, with either a laminate or wood veneer overlay.

Table 6.3

There are several different ways in which worktops can be jointed. Fig 6.21 explains how post-formed worktops are fixed and jointed.

Once the worktops have been measured and cut to fit there are a number of options available. Post-formed worktops can either be jointed using grouted butt and mitre joints or metal jointing strips. When forming routed joints the worktop will require the use of connecting bolts to ensure a tight joint. Square-edged worktops can be butted directly but they will still require bolting. Then the worktop can be screwed into place.

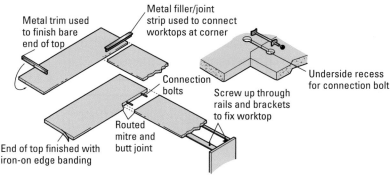

Figure 6.21 Fixing post-formed worktops

PRACTICAL TASK

2. INSTALL KITCHEN WALL AND FLOOR UNITS

OBJECTIVE

To install a range of kitchen wall and floor units according to specification.

PPE

When undertaking this task, and those that follow, ensure you select PPE appropriate to the job and site where you are working. Refer to the PPE section of Chapter 1.

PRACTICAL TIP

The majority of kitchen fitting work is completed using power tools. This is to speed up the process and also to provide a degree of accuracy and quality that only accomplished carpenters and joiners could achieve using hand tools.

TOOLS AND EQUIPMENT

Hand tools:

Tenon, panel, coping and hack saws

Claw hammer	Spirit levels
Various chisels	Smoothing plane
Screwdrivers	Block plane

Marking gauge

Try and/or combination square

G-cramps or quick release cramps (nylon faced)

Power tools:

Circular saw	HSS drill bits
Cordless drill/driver	Spur bits
Router 12 mm collet	Jig saw
Planer	110V transformer
Chop or mitre saw	Extension lead
Masonry bits	

Additional equipment:

2 × sawhorses	Worktop jig
Workmate-style bench	

STEP 1 Using the manufacturer's check sheet, ensure that all items have been delivered to site and that no components are damaged.

STEP 2 Spend some time studying all the floor plans, details and accompanying notes.

STEP 3 Carefully mark a horizontal datum line around the room at a height that will be visible throughout the build. This should be above the worktop height and below wall units.

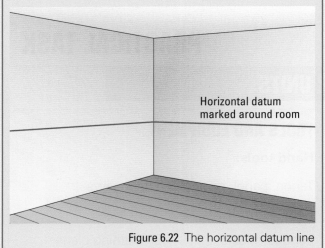

Figure 6.22 The horizontal datum line

PRACTICAL TIP

When using a level to mark a horizontal line around a room, mark one end of the level with a piece of tape or a marker pen and switch the level end for end as you proceed around the room. This reduces multiplication of error in the level.

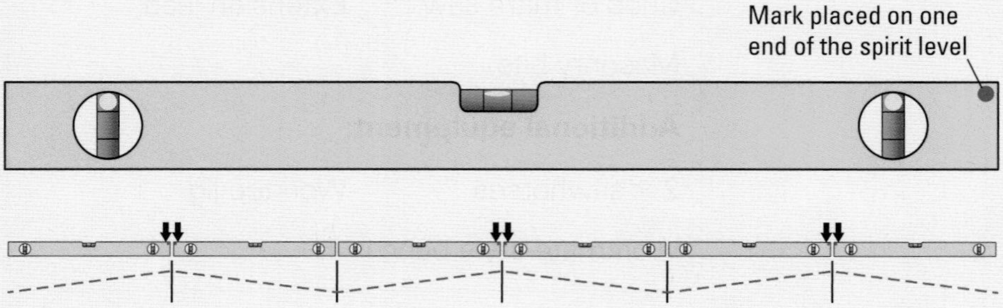

STEP 4 Measure down from datum to establish the highest point on the floor. Make a rod (a piece of 50 mm × 25 mm timber is ideal for the purpose) and stand it on the highest point on the floor and transfer the datum line onto the rod.

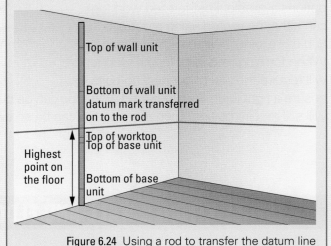

Figure 6.24 Using a rod to transfer the datum line

STEP 5 Place a mark on the rod, measured down from the datum mark to the required height of the worktop. Then add marks for the top and bottom of the base unit and the top and bottom of the wall unit on the rod.

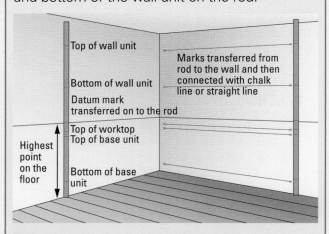

Figure 6.25 Using the rod to transfer the levels onto the wall

Mark placed on one end of the spirit level

Figure 6.23 When levelling around a room, alternate the level from end to end to reduce accumulative errors

STEP 6 Transfer the marks from the rod to the walls, where required, around the room and connect them using a straight edge or chalk line. This reduces the risk of multiplying errors when marking multiple level lines around a room.

PRACTICAL TIP

You could fasten a 45 mm × 25 mm softwood batten around the wall to the level of the underside of the worktop. This was common practice when worktops were very thin (25 mm to 30 mm), but is less common when using 40 mm plus worktops.

If you are using rigid kitchen units proceed to Step 8.

STEP 7 Carefully open the base unit packs and assemble them in accordance with the manufacturer's instructions. Avoid the temptation to open them with a utility knife as this could damage the units. Often a suggested tool list will be included with the assembly instructions.

PRACTICAL TIP

It is helpful to reuse the flat pack packaging by opening it out on the floor to protect the units during assembly.

STEP 8 Start in a corner with a base unit. Raise or lower the legs until the top is in line with the corresponding marks on the wall. The benefits of the marks for the bottom will now be clear as the unit can be levelled without you having to constantly check the top.

PRACTICAL TIP

Some systems use a corner base unit and other systems incorporate standard base units with the use of a corner post to space the units, providing the clearance for the doors or drawers.

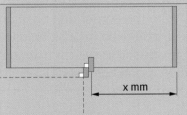

x = 400 mm for a 800 mm wide base unit
x = 500 mm for a 1,000 mm wide base unit

Figure 6.26 Fixing different types of base unit

STEP 9 With the unit located, check the measurement from the wall to the front of the unit to ensure that the worktop can cover this distance plus the thickness of a door or drawer plus the recommended overhang. If not, the units will require scribing back to the wall.

Measure from wall to the front of the unit to ensure the worktop can cover the drawer or door plus overhang

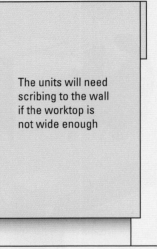

The units will need scribing to the wall if the worktop is not wide enough

Figure 6.27 Checking the measurements from the wall

STEP 10 With the corner unit positioned and levelled, it can be fixed in position using the brackets provided. This will be fixed onto the timber batten if one has been used.

PRACTICAL TIP

It will be necessary to notch the base units to allow them to sit around the battens, where used.

STEP 11 Continue fixing base units, levelling and fixing as before, scribing back to the wall as required. Subsequent units should be fixed to each other using the fixings provided. Units should be held flush with cramps to ensure a tight joint.

STEP 12 Measure carefully when leaving spaces between units for appliances allowing sufficient space for free-standing appliances to be removed, usually +15 mm.

STEP 13 If a continuous batten has not been used, fix battens to the wall between appliance spaces.

STEP 14 Fix any tall units such as larder units or fridge freezer housings; this will set the top of the wall units.

STEP 15 Carefully open the packages of the wall units and assemble them as for base units.

STEP 16 Wall units are supplied with two-part wall brackets. Fix the first part to the wall following the manufacturer's instructions. Often a template is provided.

STEP 17 Hang the wall units on to the brackets and adjust the two screws found on the brackets that are located in the top internal corners of the unit. One screw adjusts the height and the other secures the unit back to the wall.

STEP 18 Continue fixing wall units as in Step 17 and fix the wall units to each other using fixings supplied.

STEP 19 Ensure all base and wall units are level, plumb and fixed before moving on to the worktops.

PRACTICAL TASK

3. INSTALL KITCHEN WORKTOPS

OBJECTIVE

To fix laminated post-formed worktops using a butt and mitred joint.

This practical task takes you through the preferred method for quality work, but remember that jointing strips offer a cheaper alternative.

❋ Refer to the practical task *Install kitchen and wall units* for details of the tools and equipment you will need for this exercise.

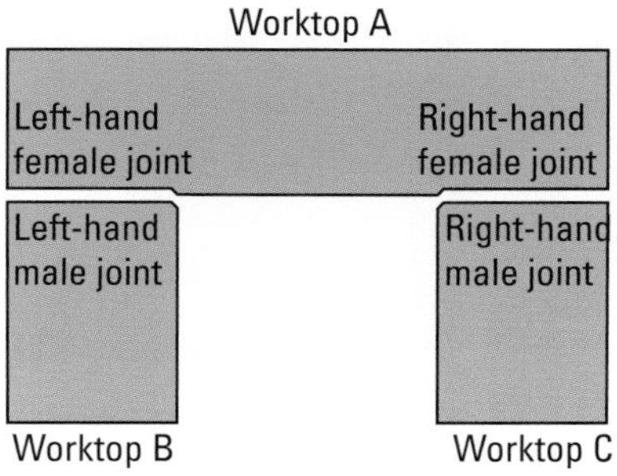

Worktop A

Left-hand female joint | Right-hand female joint

Left-hand male joint | Right-hand male joint

Worktop B | Worktop C

Figure 6.28 A typical kitchen layout with 90° corners

STEP 1 Cut worktop A to length. When fitting a worktop between two walls it may require scribing on three sides. Allow some clearance to ease removal when forming the joints.

STEP 2 To form the left-hand female joint, put worktop A face down and place the worktop jig on the post-formed edge using the pegs provided to locate the jig in position.

Figure 6.29 Worktop jig in position with worktop face down

STEP 3 Secure the worktop to prevent movement. Using a router with a ½ shank fitted with a guide bush, plunge the router to a depth of 10 mm to 15 mm and move it from left to right in the jig to make the cut. This is called a pass.

Figure 6.30 Plunge the router and make a pass from left to right across the worktop

STEP 4 Make several more passes from left to right until the cut is made.

Figure 6.31 The completed cut and worktop face up

STEP 5 To form the left-hand male joint, start with worktop B face up, and position the worktop jig using the pegs.

Figure 6.32 Make several passes until the left-hand male cut is made

STEP 6 Secure the worktop.

STEP 7 Make several passes with the router from left to right in the jig until the cut is made.

Figure 6.33 Left-hand male cut complete

STEP 9 With worktop A placed face down, position the jig to form the bolt slots and make the cuts.

STEP 10 Repeat Steps 2 to 9 for worktop C. However, worktop A will be face up to form the right-hand female joint and worktop C will be face down. This allows the router to be moved from left to right to make each cut, preventing any breakout on the post-formed edge.

STEP 8 With worktop B face down, reposition the jig to form the bolt slots. Again make several passes to achieve the required depth. (This will be about two-thirds of the depth of the worktop.)

Figure 6.34 Reposition the jig to form worktop connector bolt slots

Figure 6.35 Bolt slots after the first pass

STEP 11 Position worktop A and fix it to the base units.

STEP 12 Dry-fit worktops B and C and ensure there is a tight fit on the joints. Check that the front edge is parallel to the face of the base units.

STEP 13 Using a biscuit jointer, form slots for biscuit joints. These are set into the edge of the worktop to keep the upper surface perfectly in line. Three is sufficient for a 600 mm worktop. They should be spaced evenly across the width of the worktop.

STEP 14 Apply silicone sealant to the joints before final positioning and tightening of the bolts. Do not try to clean excess silicone while it is still wet – allow it to cure then peel it off.

PRACTICAL TIP

Sealants are available that match the colour of the worktops, making the joint almost invisible when complete.

PRACTICAL TIP

The biscuits prevent the worktop joint from slipping vertically as the bolts are tightened.

Figure 6.36 Worktop connectors in place

PRACTICAL TASK

4. INSTALL KITCHEN WORKTOPS TO HOUSE HOBS AND SINKS

OBJECTIVE

To fix laminated post-formed worktops to house hobs and sinks.

Worktops will need to be cut to house hobs and sinks. Templates are often provided with the product and these should be used when available. When laying out the shape on the worktop, use masking tape to provide a surface on which to mark the lines. The procedure is the same for both sinks and hobs. These housings need to be sealed to prevent the ingress of moisture.

Refer to the practical exercise *Install kitchen and wall units* for details of the tools and equipment you will need for this exercise.

STEP 1 Place the worktop on two sawhorses.

STEP 2 Lay out the shape of the cut onto the face of the worktop. Mask it using decorators' masking tape.

STEP 3 Drill holes on all four corners of the cut-out. These should be large enough to allow a jig saw blade to pass through.

STEP 4 Mask the bottom of the metal base using tape; this prevents the base from scratching the surface of the worktop. Fit the jig saw with a down-cutting blade.

STEP 5 Cut between the holes, supporting the cut-out throughout to prevent waste material dropping and breaking the face.

PRACTICAL TASK

5. FIT PLINTHS TO KITCHEN BASE UNITS

OBJECTIVE

To fit plinths to the legs of the base units.

Refer to the practical exercise *Install kitchen and wall units* for details of the tools and equipment you will need for this exercise.

PRACTICAL TIP

U-shaped brackets are screwed to the back of the plinth and they snap-fit onto the legs.

STEP 1 Measure the plinth length.

STEP 2 The plinth may need scribing if the floor is uneven or out of level.

STEP 3 Lay the plinth in front of the base units, ensuring that the end is line with its final position. Mark the centres of the brackets and screw them in place.

STEP 4 Push the plinth onto the base unit legs, ensuring they snap into place.

STEP 5 Repeat process around rest of the base units.

PRACTICAL TIP

When two plinths meet on an internal corner a spring clip is used to join the two plinths together; alternatively the corner could be fixed together using connector blocks, when the corner would then have to be installed as one piece.

6. FIT A CORNICE TO WALL UNITS

OBJECTIVE

To fit a cornice to a wall unit for a decorative finish.

A cornice and pelmet or a lighting rail provides a decorative finish to the top and bottom of the wall units. They also tie the units together, giving them continuity. The fixing procedure is similar for both.

Refer to the practical exercise *Install kitchen and wall units* for details of the tools and equipment you will need for this exercise.

PRACTICAL TIP

The cornice steps out over the edge of the wall units to clear the doors. This distance will be specified in the kitchen manufacturer's instructions or is sometimes indicated by a line or groove on the cornice. This will set the distance to the back of the cornice on the top of the unit. This is known as a margin.

PRACTICAL TIP

The cornice can be fixed to the units as individual pieces or glued at the mitres and offered up in sections. Mitres should be held with Mitre Mate or similar wood glue.

STEP 1 Mark the margin on to the top of the wall units. Where the margin lines cross will denote the short point of all external mitres and the long point of internal mitres.

STEP 2 Measure the first length of cornice then cut the mitre using a chop saw. Cut from the face to avoid breakout.

STEP 3 Cut the second piece of cornice and test the fit of the mitre.

STEP 4 Fix the cornice to the top of the wall units, holding the back edge to the margin using the screws specified in the installation instructions.

7. FIT DOORS TO KITCHEN UNITS

OBJECTIVE

To fit doors onto complete kitchen units.

Refer to the practical exercise *Install kitchen and wall units* for details of the tools and equipment you will need for this exercise.

STEP 1 The unit doors should be carefully unpacked.

STEP 2 Fit the hinges. This is best carried out with a hand screwdriver as the screws are very short and can easily strip their pilot holes. If a cordless screwdriver must be used, ensure the torque setting is correctly adjusted.

STEP 3 Fit the hinge mounting to the inside front edge of each unit.

STEP 4 Hang the door and adjust height, alignment, clearance and closing action with the screws on the inside of the hinge. Consult the manufacturer's instructions to ensure a good fit.

CASE STUDY

South Tyneside Council's
Housing Company

Learning on the job

Glen Campbell is a team leader at South Tyneside Homes:

'When I came out of school, I had 8 or 9 GCSEs at A to C. But something my father said always stuck with me: "If you've got a trade behind you, you've always got something to fall back on." So I started applying for apprenticeships before I got my results, and applied to the South Tyneside Council (as South Tyneside Homes used to be). As part of the apprenticeship, I completed my NVQ Levels 2 and 3 over the three-year period.

For my day-to-day work as apprentice, I was based at the maintenance depot. I would undertake anything from replacing a door handle, fixing a window, putting a new kitchen in, hanging doors, putting down new floors… all aspects really. In my first year, it was basically repairs and learning my hand skills.

Colleagues would show you how to do new things, and as long as you were receptive, you'd learn. A lot of young lads these days come in and say "I know it all already". But the reality is that they haven't done it on a live site yet.

After my first year, I was lucky enough to be chosen to come onto the capital or major works, where you're on new sites, building bungalows and new roofs from scratch, all the way through the process.'

INSTALLING SIDE HUNG DOORS AND ASSOCIATED IRONMONGERY

Door hanging makes up a large part of the carpenter and joiner's second fix work. Different types of door have different functions. There are two distinct door types:

* Internal   * External.

The design and detail of each door is in addition to its basic function.

All doors need to be easy to open and to be durable. But they may need to perform other functions, such as the following:

* Security – these may need to be solid, reinforced and have ironmongery fittings, such as security latches and bolts.

* Weatherproofing – they need to be able to provide at least the same amount of weather protection as the outside of the house. Openings are always a weak point, so their construction and fitting is very important.

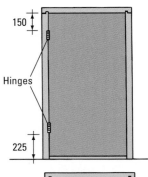

150

Hinges

225

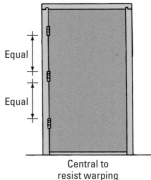

Equal

Equal

Central to
resist warping

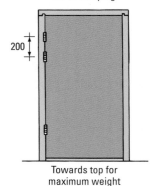

200

Towards top for
maximum weight
capacity

Hinges in
line with
rails

Figure 6.37 Hinge positioning

* Fire resistance – these may need to be able to separate parts of a building. They also need to provide an easy way to escape from the building. Some of these doors may be shut for most of the time, if not all of the time, and only used in an emergency.

* Sound or thermal insulation – these doors need to separate parts of the building. They can contribute to cutting out noise and commonly help to prevent draughts and loss of heat.

* Privacy – inside a normal domestic dwelling, doors are there to ensure privacy even if they are partly open. This means that most internal doors should cover the room.

* Ease of use – the weight and the ironmongery of a door is important. The door needs to match the traffic that will pass through it.

* Durable – most doors will be opened and closed on numerous occasions each day. They may be misused and slammed. The door and its ironmongery should be fit for purpose.

Different types of side hung door

Internal doors tend to be hung on one-and-a-half pairs (three) of 75mm hinges. These doors are relatively lightweight. Sometimes a pair will be specified (see the top diagram in Fig 6.38).

External doors are fixed on three 100mm hinges. This is also the case for one-hour fire doors. Half-hour fire doors are also usually fixed with 3 × 100mm hinges. Architects will usually state in the specification the number of hinges to be used in each situation.

The diagram shows the typical hinge positions for different types of doors.

Types of ironmongery

Ironmongery includes a wide variety of different types of door furniture. Some of these are common to all types of door, such as hinges, locks, latches and handles. Other doors will have different types of ironmongery, such as bolts or letter plates.

Hinges
The hinge secures the door to the frame and enables it to be opened and closed. Fig 6.38 shows the wide range of different hinges for doors.

Figure 6.38 Range of hinges for doors

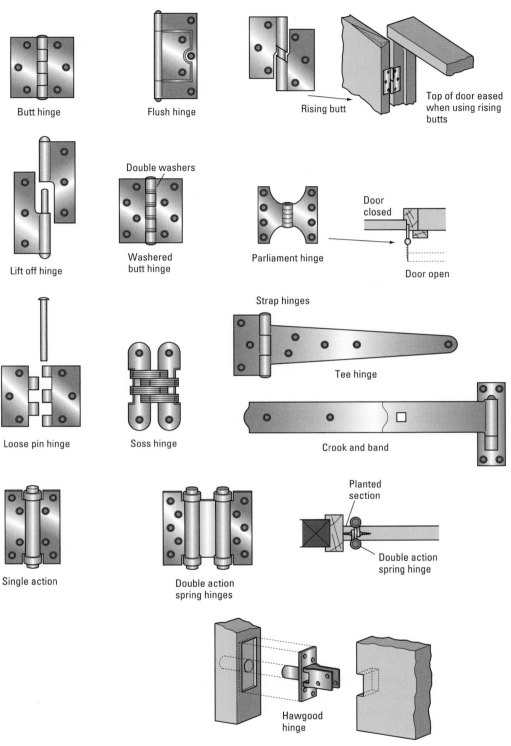

As far as hinges are concerned, there are three different options in terms of their material:

* Internal doors tend to have pressed, mild steel hinges. These are comparatively cheap and can be plated or finished in a variety of different ways. They are known in the trade as self-colour.

* Hardwood and external doors will usually have mild steel hinges although for high quality work, they may have stainless steel or solid brass hinges. It is not usual practice to put steel hinges on external doors due to the fact that they may rust and stain the timber.

* Heavyweight doors need strong, cast iron hinges. However cast iron hinges can be brittle.

There are several ways of fitting hinges, including half-and-half, feathering, knuckle-half in or out, etc.

Note that door size, width, height, location and weight dictate the number and sizes of hinges.

Locks and latches

Locks and latches are used to secure the door. Just how the door is secured will depend on where it is positioned and whether it is internal or external.

Table 6.4 outlines the different types of locks and latches.

Type of lock or latch	Description	
Cylinder rim night latch	These allow the door to be opened from the outside using a key or from the inside by turning a handle. Double locking versions are available and tend to be used for glazed doors. This is a security feature as the handle will not turn if the glazed pane is broken and someone tries to open it from the outside.	Figure 6.39
Mortise deadlock	These are key-operated. The more levers the more secure the lock. They are often added to doors where there is an existing cylinder rim latch.	Figure 6.40
Mortise latch	These are internal latches and do not lock. They simply hold the door in a closed position. The door is opened by using the handle.	Figure 6.41
Horizontal mortise lock or latch	This is a combination of a mortise deadlock and latch. This lock is used on external doors. It usually has 5 levers and is insurance-rated in the same way as a vertical mortice lock latch. It differs in that it is designed to be used with door knobs instead of lever handles and it gives enough clearance for fingers against the door frame. Horizontal latches are relatively rare because of their length.	Figure 6.42

Vertical mortise lock or latch	These are also known as narrow style locks or latches. They are a useful general purpose lock or latch used on external doors.	Figure 6.43
Rebated mortise lock or latch	These are used for double doors with rebated styles.	Figure 6.44
Rim deadlock	These are surface-fixed locks. They are still used in refurbishment of period buildings and heritage-type work.	Figure 6.45
Rim lock or latch	These tend to be used either for garden sheds, gates or refurbishment works. Knobs or handles operate the latch and they are surface-fixed.	Figure 6.46

Table 6.4

With each different type of lock there are a number of key accessories that add to the range of ironmongery:

* Knob sets – these are mortise latches and a pair of knob handles that can be locked with a key.

* Escutcheons – these are designed to give a neat finish to a keyhole.

* Lever furniture – most lever handles consist of a simple pair of matching handles that use the principle of leverage to turn the spindle (as an alternative to knob-style handles, which require a twisting action).

* Padlocks, hasps and staples – these are usually used on sheds and gates. The hasp is fixed to the door; the staple to the frame. When the door is closed the padlock secures the door.

Cylinders

Fig 6.47 shows how a cylinder rim night latch is fitted.

The manufacturer will tend to supply a template showing the recommended height and position of the centre of the cylinder hole. The cylinder hole needs to be drilled out and then the cylinder is passed through the hole and secured to a mounting plate.

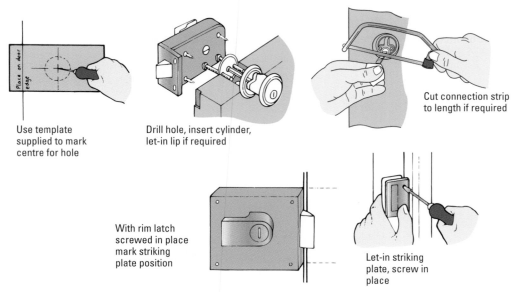

Use template supplied to mark centre for hole

Drill hole, insert cylinder, let-in lip if required

Cut connection strip to length if required

With rim latch screwed in place mark striking plate position

Let-in striking plate, screw in place

Figure 6.47 Fixing a cylinder rim night latch

Most of the latches are designed to fit doors of varying thicknesses and it is necessary to use a hack saw to trim the flat connection strip to fit. The case is then fixed over the mounting plate and marked out. It can then be secured and checked for smooth operation.

The cylinder rim latch case will show where the striking plate needs to be positioned when the door is closed. This will need to be chiselled out of the door frame and then the striking plate can be secured in place with wood screws.

Bolts and chains

Bolts and chains are additional security devices. There are a number of different types, which are covered in Table 6.5.

Types of bolt or chain	Description and use
Barrel and tower bolts	These are used on external doors and gates, with one bolt at the top and one at the bottom.
Flush bolts	These are recessed into the wood and can be used for external doors, particularly double doors and French windows. There is usually one at the top and one at the bottom.
Mortise bolts	These are key operated dead bolts that are fixed into the edge of the door at the top and the bottom.
Panic bolts	These have push bars that disengage the bolts. They tend to be used for emergency exit doors.
Hinge bolts	These are designed to stop doors from being forced off their hinges. They tend to be used on doors that open inwards to prevent the hinge side being crow-barred out against the rebate.
Door holders	These are foot operated stops that keep doors in an open position.
Security chains	These have slides fixed to the doors and chains fixed to the frames. When they are in operation the door can only be opened a limited amount.

Table 6.5

The position of locks, latches, bolts and chains can be seen in Fig 6.48.

Letter plates

Letter plates can be fairly straightforward designs and are essentially a framed opening in the door with or without a hinged cover. Some may have integral draught-proofing in the form of plastic fibre strands.

The position of letter plates is largely dependent on the design of the door. They can be fitted into the middle or bottom rail, and vertically into some styles of door. In fact letter plates can be purchased with a pre-drilled hole for a cylinder rim latch. This has to be fitted to accommodate the lock. Doors hung in wing lights or porch frames will often have the letter plate mounted in the frame rather than the door. This has the advantage of not weakening the door.

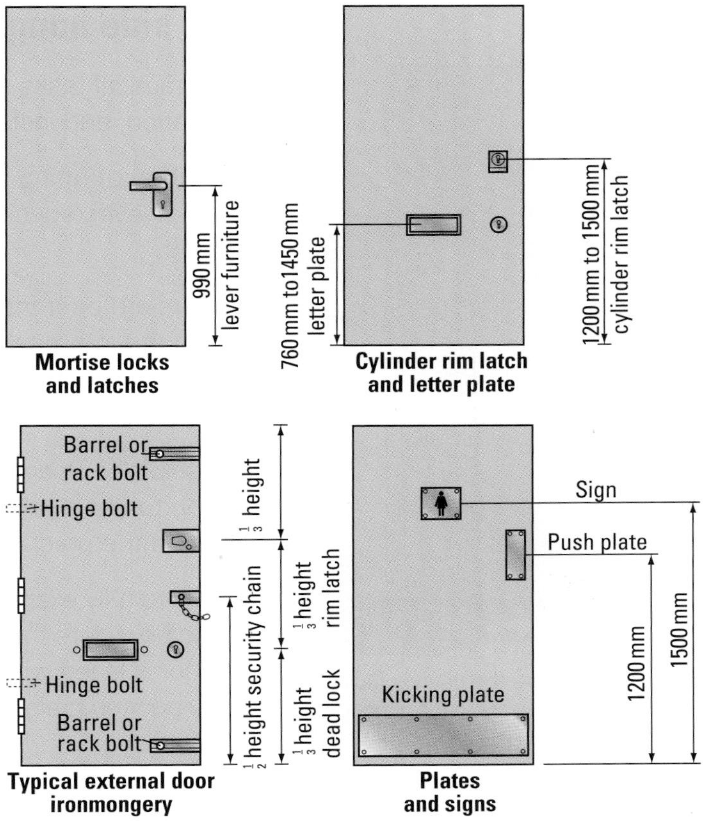

Letter plates differ in their manufacture in terms of overlap and spring positions. Templates are often provided and should be used to ensure a good fit. If numerous letter plates of the same design are to be fitted, for instance on a housing development, it is a good idea to make a template for use with a router to get a consistently good finish and to speed up installation.

Mortise locks and latches — 990 mm lever furniture

Cylinder rim latch and letter plate — 760 mm to 1450 mm letter plate; 1200 mm to 1500 mm cylinder rim latch

Typical external door ironmongery — Barrel or rack bolt; Hinge bolt; Hinge bolt; Barrel or rack bolt; $\frac{1}{2}$ height security chain; $\frac{1}{3}$ height dead lock; $\frac{1}{3}$ height rim latch; $\frac{1}{3}$ height

Plates and signs — Sign; Push plate; Kicking plate; 1200 mm; 1500 mm

Figure 6.48 Fixing heights of locks, latches, bolts and chains

Door closers

These are not so common in domestic dwellings, but are used widely in commercial properties. They tend to be fitted to heavy doors to provide a self-closing action. They will also allow the door to be held open.

The door closer can be adjusted to determine how fast the door closes. Others are fitted with temperature sensors, which will automatically close the door if there is a fire. Other door closers have a delayed closing action; some will snap shut when nearly closed to ensure that the latch is engaged; others will have a safety mechanism to prevent them from slamming against walls when opened.

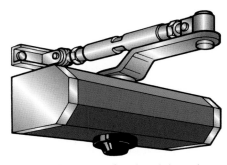

Figure 6.49 Overhead door closer

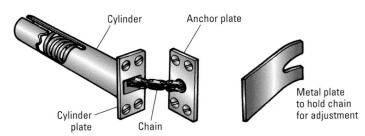

Cylinder; Anchor plate; Cylinder plate; Chain; Metal plate to hold chain for adjustment

Figure 6.50 Concealed door closer

Fixing side hung doors

These practical tasks show how to hang a variety of doors to a given specification, and install associated ironmongery.

The method of fitting a side hung door can be applied to most types of doors; however, special regulations govern the installation and fixing of fire doors.

Fire doors are best fitted into their frames in a workshop environment, installing the intumescent strips and the smoke seal after fitting is complete. They can then be delivered to site in a set to be fitted into pre-prepared openings.

It is possible to fit and hang doors with hand tools but it is now more common to use a mixture of hand and power tools where appropriate, as listed in the practical task below.

You need to fully examine the door to determine its best face, which needs to be where the surface of the door is seen on the most occasions. If the door is faced with a timber surface, examine the grain and ensure that any pointed grain faces upwards.

Figure 6.51 Ensure the grain is facing in the right direction

PRACTICAL TASK

8. FIT A SIDE HUNG DOOR

OBJECTIVE

To fit a door into a door frame.

This list applies to all the practical tasks that follow.

PPE

Ensure you select PPE appropriate to the job and site where you are working. Refer to the PPE section of Chapter 1.

TOOLS AND EQUIPMENT

Hand tools:

Claw hammer	Bradawl
Mallet	Combination square
Screwdriver	Marking gauge or butt gauge
Jack or fore plane	Panel saw
Smoothing plane	Tape measure or rule
Chisels	Utility knife

Carpenters brace with various size auger bits

Power tools:

Cordless drill/driver with bits

Portable powered plane

Portable powered circular saw

Jig saw

Portable powered router with an assortment of cutters

Transformer   Extension lead

In addition to the tool list the following may be advantageous:

2 × saw horses/stools   Proprietary hinge, letter plate or mortise lock templates

STEP 1 Whenever drawings and door schedules are available check them to confirm size, type, style, position and hang of the door.

PRACTICAL TIP

If drawings or schedules are not available, the door should be hung, whenever possible, in a manner that gives maximum privacy to the largest part of the room. This is known as 'covering the room'. However, in modern construction this is not always possible, as the position of light switches and sockets would obviously affect the hanging side, and with ever-decreasing room space, the area needs to be maximised.

STEP 2 Check the overall size of the door before doing any cutting. Many doors are supplied with lugs or horns from the manufacturing process left on for protection. If they are still on the door you are fitting, place the door onto two saw horses, mark a square face and edge across each of the horns of the door with a try or combination square and a sharp pencil, then remove the horns using a panel saw, keeping the angle of the saw low to minimise break out on the underside.

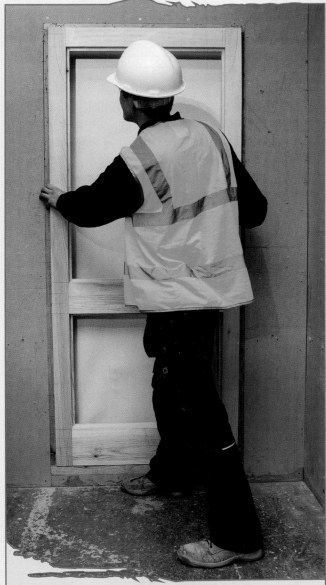

Figure 6.53 Remove just enough to allow the door to fit in the frame

Figure 6.52 Removing the horns

STEP 3 Once the horns are removed, shoot (plane) the edges and the ends of the door, removing the minimum amount of material to allow the door to sit inside the frame.

STEP 4 Using a jack plane, shoot the hanging side of the door to fit the jamb.

Figure 6.54 Shooting the hanging side

STEP 5 Shoot the top edge of the door to fit the head if required.

STEP 6 Wedge the door tight to the head of the frame and deduct top and bottom clearance from the bottom of the door. This should be around 2 to 3 mm at the top. The clearance at the bottom will depend on floor coverings, the use of storm guard cills and anything else that takes up space beneath the door.

Figure 6.55 Setting clearance at the bottom of the door

Figure 6.56 Setting clearance at the top of the door

PRACTICAL TIP

When hanging doors in frames with water bars they will require a rebate forming at the bottom of the door.

STEP 7 Shoot the closing stile or edge of the door until it is parallel and the gap is twice the required clearance. This should be done with the door wedged tight in the frame to the hinge side and the head.

STEP 8 Shoot a slight leading edge on the hinge and closing stiles; this should be minimal to avoid problems when fitting the lock. Remove the arris (sharp edge of the timber) from all edges of the door with a smoothing, block plane or glass paper.

Figure 6.57 Checking the leading edge with a square

Figure 6.58 Plane at an angle

PRACTICAL TIP

Shooting a leading edge on the hinge side of the door ensures that the hinge never fully closes, reducing the chance of the door being hinge-bound. This is where there is insufficient clearance for the hinges or the door against the frame or lining. Hinge-bound is a term that describes the effect of the hinge side of the door springing against either the screw heads or the back of the rebate. It is a result of poor workmanship, often because the hinges have been cut too deep.

STEP 9 Place the door back in the opening. Put a suitable spacer (such as the blade from a combination square) between the head of the door and the frame; wedge the door tight to the hinge side and the spacer at the head. Mark the positions of the top of the hinges on both the door and the frame.

PRACTICAL TASK

9. FIT HINGES TO A DOOR

OBJECTIVE

To prepare the hinge pockets of a door.

PRACTICAL TIP

Take the position of the hinges from the door schedule, if there is one. Often there is not one and in these circumstances the industry-accepted standard for the UK is 150 mm down from the top of the door and 225 mm from the bottom, with the third hinge in the centre of the two. Hinges are normally described as being in pairs, which simply means that there are two matching hinges.

3 × (1½ pairs of) 100 mm butt hinges are needed for external doors

3 × (1½ pairs of) 75 mm butt hinges are needed for internal doors

STEP 1 Having marked the top of all three hinges (Step 9 in the task above) square these lines onto the edge of the door. Offer a hinge up to the marks and place a pencil mark for the bottom of the hinge on the edge of the door. Do not draw around the hinge. Square these marks across the door edge and back onto the face.

Figure 6.59 Marking the position of the hinges

PRACTICAL TIP

It is good practice to use cut lines to provide an accurate position for paring the hinge.

STEP 2 Set a marking gauge or butt gauge to the hinge leaf width. Gauge leaf width along the door edge between the marks.

Figure 6.60 Gauging the leaf width

STEP 3 Set a marking or butt gauge to the hinge leaf thickness and mark between the lines on the face of the door.

Figure 6.61 Gauging the leaf depth

STEP 4 Mark the hinge leaf height onto the frame or lining in the same way.

STEP 5 While the gauge is set for leaf thickness, gauge between the marks on the face of the frame or lining.

STEP 6 Chop the ends of the hinge pockets.

Figure 6.62 Chopping the ends of the hinge pockets

STEP 7 Feather across the grain to the required depth.

Figure 6.63 Feathering to the required depth

STEP 8 Lightly chop the back edge of the hinge pocket with a wide chisel.

Figure 6.64 Using a chisel to chop the back edge

STEP 9 Pare across the grain to the back of the hinge pocket and then work the holes for the hinge with a bradawl, using the hinge as a guide.

Figure 6.65 Paring out the waste

Figure 6.66 The hinge pockets ready for installation

STEP 10 Repeat Steps 6 to 9 for hinge pockets on the frame or lining.

Figure 6.67 Using the bradawl to mark the holes

PRACTICAL TASK

10. PREPARE A DOOR'S HINGE POCKETS

OBJECTIVE

To prepare the hinge pockets on the door frame.

STEP 1 Mark the hinge leaf height onto the frame or lining in the same way as for the door at the end of Task 8.

STEP 2 While the gauge is set for leaf thickness, gauge between the marks on the face of the frame or lining.

STEP 3 Measure the distance from the back of the hinge pocket on the door to the back edge of the door. Add approximately 2 mm to this measurement and call this 'x'.

STEP 4 Measure out 'x' from the back of the rebate on the frame and mark with a pencil. Set a gauge from the face of the frame to this mark. Gauge between the hinge leaf height marks.

PRACTICAL TIP

Many carpenter/joiners modify marking gauges by shortening the end of the gauge so that it will fit into the rebate on a door frame. Butt gauges will do this job without modification; however they are not as commonly used these days.

STEP 5 Chop the ends of the hinge pockets.

Figure 6.68 Chopping the ends of the hinge pockets

STEP 6 Feather across the grain to the required depth

STEP 7 Lightly chop the back edge of the hinge pocket with a wide chisel.

STEP 8 Pare across the grain to the back of the hinge pocket.

Figure 6.69 Paring across the grain to remove the waste

STEP 9 Repeat Steps 6 to 8 for hinge pockets on the frame or lining.

PRACTICAL TASK

11. HANG THE DOOR

OBJECTIVE

To fix a side hung door in position.

STEP 1 Using one of the hinges as a template, place it in each of the hinge pockets and drill a pilot hole through the lowest hole of the hinge.

STEP 2 Screw the hinges to the door.

STEP 3 Offer the door up to the frame, so that it is approximately 90° to the frame in the open position. The door should be angled back slightly to lift it, making it possible to locate the screw into the pilot hole for the top hinge. Partially drive this screw leaving the head clear of the hinge leaf.

Figure 6.70 The top pilot screw

PRACTICAL TIP

Make sure that the hinges are partially closed to ensure that the lower hinges do not get fouled behind the frame when the top hinge is located and fastened. Offset the screw to the back of the hole so that it pulls the hinge in tight to the back of the pocket.

KEY TERMS

Wind

– if something is said to be 'in wind' it is twisted. It is pronounced 'wined' as in 'wined and dined'.

STEP 4 Pull the door out to the bottom hinge pocket and secure with screws.

STEP 5 Locate the middle hinge, push the top of the door in towards the frame and drive the top screw fully home.

Figure 6.71 Fixing the bottom pilot screw to ensure the door is parallel

STEP 6 Drive one screw into the middle and swing the door into the closed position to test for fit. If they fit and clearance is good, fix all remaining screws.

Some fine tuning or secondary fitting may need to be carried out at this point.

Figure 6.72 The fitted door

TROUBLESHOOTING

Problem	Action
The door is binding against the back of the rebate.	1. Loosen each hinge in turn. 2. Ease hinge out towards front of frame. 3. Drill a new pilot hole in another hole. 4. Drive home screws.
The door is hinge-bound (it seems to be springing on the hinge side). Hinge pockets are too deep.	1. Cut suitable packings to establish an even clearance gap down the hinge side. This may not be necessary for all hinges. Veneers of a similar timber make the most suitable packings – try to avoid screw box lids, glass paper and cigarette packets. 2. Place the packing in dry and test fit. 3. If all is well, unscrew and remove the packing then apply a little PVA to the back of the packing. 4. Replace and re-fix.
The door has the correct clearance gap on all edges when closed but catches the rebate when opening and closing.	Leading edge is not sufficient; this may not be along the full length. Shoot edges where required, remembering to take the arris off. It may be necessary to remove the door.
The door is not sitting tight to the rebate on the lock side.	Either the door is in **wind** or the frame is in wind. If the gap is very small it can be solved by letting out a hinge diagonally from the gap; the centre hinge will also need letting out but by a lesser amount. This has the effect of halving the twist. If the gap is larger the rebate will need to be opened up with a rebate plane and finished into the corner with a sharp chisel.

Table 6.6

PRACTICAL TASK

12. FIT MORTICE LOCKS AND LATCHES

OBJECTIVE

To fit a mortice lock and latch into a door.

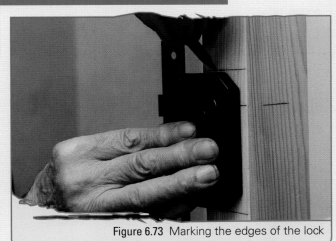

Figure 6.73 Marking the edges of the lock

STEP 1 Wedge the door open.

STEP 2 Mark the spindle height on the edge of the door. This will be approximately 990 mm; however, if the handle has to match existing furniture, take a site measurement.

STEP 3 Line the spindle hole up with the mark on the door, then mark the top and bottom edges of the lock on the edge of the door.

STEP 4 Set a marking gauge to half the thickness of the door and gauge a line between the two marks.

STEP 5 Drill a series of overlapping holes to the depth of the lock, along the centre line, using an auger that is slightly more than the thickness of the lock. The diameter of the bit should be checked after measuring the actual lock to be fitted. Locks include full fitting instructions but if these are not available, the holes should be just wider than the lock case.

Figure 6.74 Drilling overlapping holes with an auger

Figure 6.75 Drilling overlapping holes with a drill

STEP 6 Pare the sides with a firmer or bevel-edged chisel to form the sides of the mortise.

STEP 7 Square the top and bottom edges with a firmer or mortise chisel.

STEP 8 Clean debris out of the mortise. Avoid blowing in the hole as dust can often end up in your eye.

Figure 6.76 Preparing the mortise

STEP 9 Push the body of the lock into the mortise. Do not force the lock in – if it does not fit, you can remove more material from any high spots within the mortise.

Figure 6.77 The mortice with the lock

PRACTICAL TIP

With the lock sitting in the mortise, the back of the face plate should sit tight against the door edge. Some locks have a separate face plate; ensure that this is on the front of the lock.

STEP 10 You now need to form a recess that will leave the lock face plate flush with the edge of the door. There are a couple of options:

Option A Draw around the lock, feather with a chisel and remove the timber within the lines.

Figure 6.78 Feathering to the required depth

Option B Use two small screws to fasten the face plate and lock in position, then carefully cut around the face plate with a utility knife, angling the blade in slightly to keep the edge as tight as possible. Remove screws and lock, then form recess as above.

Figure 6.79 Forming a recess, option B

STEP 11 Hold the lock against the face of the door, line it up with the face plate housing on the edge of the door and, using a bradawl, mark the centres of the spindle and the keyhole. Square these marks onto the opposite face and measure the same distance in from the edge of the door, remembering to make the allowance for the leading edge.

Figure 6.80 Marking the position of the spindle and keyhole

STEP 12 Drill a 16mm hole for the spindle. Make sure you drill from both sides to avoid breakout.

STEP 13 Drill a 10mm hole for the keyhole. Then form the keyhole by cutting down with a pad saw and finishing it off with a chisel, again working from both sides of the door. Alternatively, drill a second hole (6mm in diameter) just below the first and clean out with a sharp bevel edged chisel.

Figure 6.81 Making a hole with an auger

Figure 6.82 Making a hole with a drill

PRACTICAL TIP

Never drill one large hole for the keyhole. The hole should be formed as described in Step 13 to provide a guide for the key, and to maximise the hole's strength by retaining as much material as possible next to the lock.

STEP 14 Insert the lock along with the face plate. Check that the spindle hole and the keyhole line up when the face plate is flush. Check the operation of the key in the lock. Fix using the screws provided.

Figure 6.83 Checking the lock

STEP 15 With the dead bolt out, close the door against the frame and mark the positions of the bolt and latch. Use a combination square to mark lines on to the face of the frame.

STEP 16 Measure the distance from the outside of the door to the square face of the latch. Now measure out this distance from the back of the rebate, adding 2 mm, and mark between the latch marks on the frame.

STEP 17 Hold the striking plate in position, lining up the front edge of the latch mortice with the mark on the frame. Ensure that the mortises line up with the corresponding marks for the height of latch and dead bolt. Screw the face plate in position and cut around it using a utility knife. Mark the mortises using a pencil.

STEP 18 Using a sharp chisel, chop out the mortises for the bolt and latch, and then chisel out the recess to the striking plate depth. The striking plate has a lead in to the mortises, and to allow this to sit flush, the corner must be taken off the frame across its width. Fix using the screws provided.

Figure 6.84 Positioning the striking plate

STEP 19 Fit the specified lever handle set or door furniture.

STEP 20 Check that the latch engages smoothly and that the dead bolt can be locked easily from inside and outside with all the keys provided. Check that the door does not rattle in the frame.

Figure 6.85 Door with the latch and handle fitted

PRACTICAL TASK

13. FIT CYLINDER RIM LOCKS

OBJECTIVE

To fit a cylinder rim lock into a door.

STEP 1 Wedge the door open.

STEP 2 Read the instructions supplied with the lock. Mark the centre of the cylinder hole at the required height, using the paper template supplied as part of the instructions.

STEP 3 Drill a 32mm diameter hole with a suitable drill bit such as a spade bit or auger. Drill from both sides to avoid breakout and to keep the hole square to the edge of the door.

STEP 4 Place the cylinder in the hole from the outside of the door and secure it using the mounting plate and the two long machine screws provided.

Figure 6.86 The secured cylinder

STEP 5 Place the key in the cylinder and keep the key slot vertical as the screws are tightened. These screws may need shortening with a hack saw. Take care as the strength of the lock is reduced if these machine screws are only just long enough. Cutting points into the screw are incorporated into better quality versions to avoid cross threading.

STEP 6 When the cylinder is suitably aligned by checking the operating parts of the lock, secure the mounting plate to the back of the door using the wood screws provided.

STEP 7 The flat strip that connects the cylinder to the lock case will need shortening – it should project past the backplate by about 16mm. Check the installation instructions to confirm this projection.

STEP 8 Line up the arrow on the backplate of the lock casing with the slot in the thimble.

PRACTICAL TIP

Some models of cylinder lock require letting in to the edge of the door.

STEP 9 Offer the lock case over the mounting plate, making sure the connecting strip is engaged in the slot.

Figure 6.87 Ensuring the connecting strip is engaged

STEP 10 Secure the lock case to the door. Some models are secured with two machine screws into the mounting plate while others are fixed directly to the surface of the door with wood screws.

STEP 11 Check that the key and lever handle both operate the latch and that the latch can be locked off.

STEP 12 Close the door and use the lock casing to mark out the position of the keep or striking plate onto the door frame.

STEP 13 Offer the keep onto the frame between the marks. Mark the shape of the fixing plate onto the frame and let-in using a sharp chisel.

STEP 14 Fix the keep using the screws provided.

STEP 15 Check that the lock operates smoothly from both sides of the door.

Figure 6.88 The finished door

PRACTICAL TASK

14. FIT LETTER PLATES

OBJECTIVE

To fit a letter plate into a door.

STEP 1 Remove the door and place it on a pair of saw horses.

STEP 2 Mark the centre line of the plate on the face of the door. Centralise the letter plate on the centre line and draw around the outer edge.

STEP 3 Measure the size of the flap and mark this on the door, adding 2–5 millimetres for clearance.

STEP 4 Mark the bolt holes.

PRACTICAL TIP

Letter plates can be installed with the door on; however, fitting is made easier with the door off and it is worth the extra effort of doing this.

STEP 5 Drill holes for the fixing bolts. To avoid breakout, alternately drill from both sides of the door. Drill a fine pilot hole through the door before drilling the clearance hole from both sides or clamp a sacrificial piece of wood to the back of the door.

STEP 6 Drill holes at each corner of the flap cut out to aid the use of a jig saw if you are using one. If you are using a purpose-made letterbox jig and a router, this stage is unnecessary.

STEP 7 Use a jig saw to cut between the holes, scoring across the cross grain with a utility knife to minimise breakout on the upper face.

STEP 8 Clean up the inside of the opening with glass paper and remove the arris.

STEP 9 Secure the letter plate with the bolts provided, and ensure that the plate is parallel to the rail in which it is housed.

STEP 10 Check operation and adjust if necessary.

Fire doors are fitted in much the same way as any other side hung door, with a number of additional considerations. All fire doors are fire-rated and this can be found in one of two places on the door – either a label on the top or on a plug in the edge, which is colour-coded, denoting its particular fire rating.

Fire doors are best fitted into their frames in a workshop environment, installing the intumescent strips and the smoke seal after fitting is complete. They can then be delivered to site in a set to be fitted into pre-prepared openings.

PRACTICAL TASK

15. FIT FIRE DOORS

OBJECTIVE

To fit a fire door.

Consult the specification, door schedule and drawings before starting work.

STEP 1 Remove the intumescent strip from the door.

STEP 2 Fit the door in the normal way, taking particular care with clearance gaps. These should be within the parameters of the specification to allow the correct functioning of intumescent strips and smoke seals.

STEP 3 Using a router, re-house the strips or seals back to their original depth.

STEP 4 In some instances the frame will require routing out to take the strip. This can also be done in situ using a router for the bulk of the work, and finishing to the corners and the bottom of the casing legs with a sharp chisel.

STEP 5 Hang the door in the normal way. Do not remove the label or plug from the door that states its fire rating as this needs to be visible for fire regulation inspections.

PRACTICAL TIP

Some frames have double intumescent strips that are parallel to each other in the frame. This ensures that there is a continuous seal when the hinge is chopped through the first strip.

INSTALLING INTERNAL TIMBER MOULDINGS

Timber mouldings are ornamental contours and shapes that are used for decorative purposes, named after their profile, as can be seen in Fig 6.89. Timber internal mouldings use geometric shapes based on Grecian or Roman designs. Grecian designs are based on elliptical shapes and Roman on circles. They are used to finish off openings where the plaster meets the frames or floors.

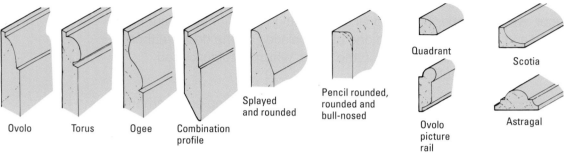

Figure 6.89 Mouldings

Mouldings such as these can be fitted in a variety of different places. They are installed at second fix stage and are mainly for internal use. In a typical room timber mouldings can be used for the skirting board or architrave and for additional decorative features such as dado or picture rails.

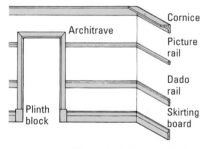

Figure 6.90 Types of trim

Types and sizes of mouldings

The different types of mouldings are shaped and sized for particular uses, as can be seen in Fig 6.90.

Each of these different types of trim has a particular purpose, as can be seen in Fig 6.91. Most of these functions are to cover up gaps or to protect plaster work.

REED TIP

It is useful to have a driving licence, especially if you want to be taken on permanently. You may need to drive vans and trucks.

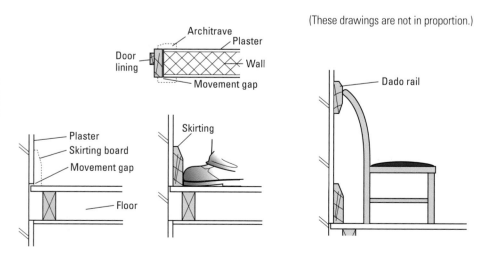

Figure 6.91 Covering gaps and protecting plaster work

The main types of timber mouldings or trim are outlined in Table 6.7.

Timber moulding or trim	Use and position
Skirting	This is fixed at the bottom of the wall. It covers the joint between the floor and the wall. It is there to also protect the plaster from kicks and scrapes.
Architrave	This is a timber moulding that is placed around either door or window openings. It serves a double purpose as it covers the joint between the wall and the timber and also hides any gaps due to shrinkage.
Dado rail	This is traditionally placed at a height of a tall chair back. This means it is usually around 1 m from the floor. Sometimes it is referred to as a chair rail.
Picture rail	This is usually positioned around 1.8 to 2.1 m from the floor. It has a special shape designed to allow clips to be hooked over it to hold picture frame wire.
Plinth block	This is fixed at the base of an architrave. It is usually wider and taller than the skirting that abuts it because on deeper section skirtings the thickness is increased, and is often thicker than the architrave it would usually sit against.
Cornice	These are mouldings positioned at the junction of the wall and ceiling. They can be made from timber but are usually made out of plaster or polystyrene.

Table 6.7

Skirting boards and architraves provide a finish to the edges of plastering where it meets the floor and around frames and linings, concealing any gaps caused through shrinkage and protecting the plaster from knocks and bumps.

Dado rails, sometimes known as chair rails, again provide protection for the plasterwork and split the height of the room in older properties, providing a finish for wallpapers or contrasting finishes above and below. Picture rails were installed in the past to provide a means of hanging pictures without damaging the walls.

The shape and size of the timber being used for these various purposes will depend on the specification and styling of the room. They can be made from hardwood, softwood or MDF. Most of them are supplied on site ready for installation.

Fixing, mitring and scribing mouldings

The following practical tasks show how to fix, mitre and scribe mouldings.

The process is shown for installing skirting boards but should be followed for installing dado and picture rails as well.

It is usual to deal with the architrave first. Once the architrave has been fitted along with a plinth block if this is being used, the skirting is then cut and fixed. After this any other decorative features are then measured up, cut and fixed.

16. INSTALL ARCHITRAVES

OBJECTIVE

To fix architraves around a door.

Architraves are fixed after the doors have been hung; this makes it easier by allowing a marking gauge to be run down the face of the frame or lining, to mark out the hinge pockets. Unless a plinth block is to be used, architraves should be at least as thick as the skirting board.

PPE

In this task, and those that follow, ensure you select PPE appropriate to the job and site where you are working. Refer to the PPE section of Chapter 1.

TOOLS AND EQUIPMENT

Hand tools:		**Power tools:**	
Combination square	Mitre box	Hammer drill	Plug cutter
Tape measure	Pair of compasses	Cordless screwdriver	Chop saw
Pencil	Block plane	HSS bits	Nail gun
Tenon saw	Panel saw	Countersink	110V transformer
Claw hammer	Coping saw	Counter bore	Extension lead
Nail punch	Screwdriver		
Sliding bevel			

These tools and equipment apply to all the practical tasks that follow.

STEP 1 Architraves should be fixed to a margin – a line around the frame or lining that is set back from the front edge by approximately 6 mm. Use a combination square to transfer a pencil line onto both legs and the head; this line should meet in the corners.

PRACTICAL TIP

If you are right-handed, you will probably find it easier to work from left to right and vice versa if you are left-handed.

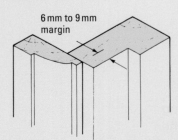

6 mm to 9 mm margin

Figure 6.92 Margin to architraves should be approx. 6 mm

STEP 2 Cut a square end on a length of architrave. Stand this cut on the floor and offer the architrave up to the frame (you may need to tack this in place temporarily) with the thinnest edge touching the margin line. Mark the architrave where the vertical and horizontal margin lines meet.

STEP 3 The mark on the architrave represents the short point of the mitre joint. Place the architrave in a mitre box and cut it with a tenon saw or chop saw if available and if you have been sufficiently trained and are competent in its operation. Alternatively, mark a 45° line across the face of the architrave and cut it freehand, undercutting slightly to ensure a tight fit on the face.

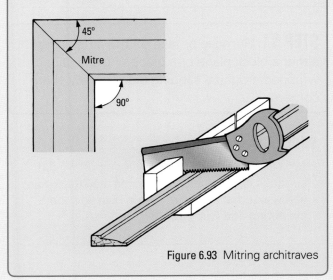

Figure 6.93 Mitring architraves

PRACTICAL TIP

When multiple sets of architrave are to be fitted this operation is best completed with a portable powered chop saw, but you must be careful where there are varying floor levels.

STEP 4 Place the architrave leg against the margin, check its length and temporarily fix it using oval nails. Do not use round-headed nails such as lost heads or pins as these do not close around the grain when punched below the surface. Try to disguise the fixings by nailing through quirks (deep penetrations already in the moulded surface) where possible.

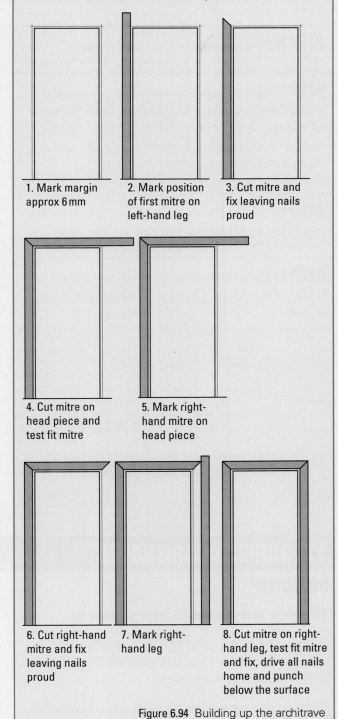

1. Mark margin approx 6 mm

2. Mark position of first mitre on left-hand leg

3. Cut mitre and fix leaving nails proud

4. Cut mitre on head piece and test fit mitre

5. Mark right-hand mitre on head piece

6. Cut right-hand mitre and fix leaving nails proud

7. Mark right-hand leg

8. Cut mitre on right-hand leg, test fit mitre and fix, drive all nails home and punch below the surface

Figure 6.94 Building up the architrave

STEP 5 Cut a 45° angle on the head piece and offer it up to the architrave that has already been fixed. Check the fit – if the joint is tight, mark the other end of the head piece at the intersection of the margin and cut a 45° angle. If there is a gap, the headpiece will need planing with a block plane until the joint fits tightly, taking care to support the external edge.

STEP 6 Temporarily fix the head piece.

STEP 7 Cut a square end on the second leg and offer it up to the head piece, with the back of the leg facing outwards, and mark the long point from the head. Transfer this mark on to the face and cut as before.

STEP 8 Check the fit as in Step 5 and trim the leg to fit, and then fix it as before.

STEP 9 Drive all the nails home and punch below the surface of the architrave with a nail punch.

STEP 10 Check that the joints are flush on the face and pin the mitres through the head into the legs. A small amount of PVA adhesive can be applied to this mitre prior to pinning.

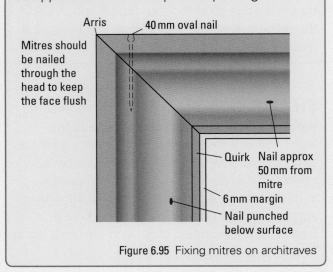

Figure 6.95 Fixing mitres on architraves

STEP 11 Remove or take off the arris using either a block plane or a piece of glass paper wrapped around a sanding block on any sharp corners.

PRACTICAL TIP

Arris is a trade term for the meeting of two flat surfaces at a corner. It describes the sharp point of the external angle. This term is used in both plastering and woodworking.

PRACTICAL TASK

17. SCRIBE ARCHITRAVES

OBJECTIVE

To scribe architraves to gain a good fit.

Architraves often need scribing where there is not enough room to accommodate the full width of the architrave in corners.

Scribing is a term used to describe the process of cutting materials to fit irregularly shaped walls or the profile of mouldings. In this instance the square edge of an architrave is scribed to a wall; we do not assume that plastering is completely flat. Scribes are also used on internal corners where two mouldings meet and one moulding is cut to fit over the other. This is because we only get single shrinkage on a scribe and double shrinkage on a mitre. On mouldings wherever possible you would use a scribe.

STEP 1 Temporarily nail the architrave to the door lining, allowing the architrave to overhang the edge. The overhang should be equal all the way down the leg.

STEP 2 Measure the overhang and add the width of the margin. Set a pair of compasses or cut a gauge block to this resulting size.

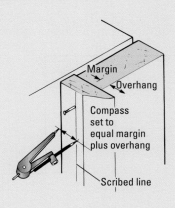

Figure 6.96 Scribing architraves

STEP 3 With the compass held against the wall, mark the line to be cut.

STEP 4 Using a panel, coping saw or a jig saw, cut along the line, undercutting slightly to ensure a tight fit on the face. Using a panel saw, remove any high points with a block plane until the required fit is obtained.

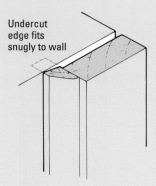

Figure 6.97 Scribing should fit snugly to the wall surface

PRACTICAL TASK

18. INSTALLING SKIRTING

OBJECTIVE

To fit and fix skirting around a room.

There are three joints used when fixing skirting:

* mitre (used on external corners)

* scribe (used on internal corners)

* heading joint (used to joint two pieces in length). It is good practice to mitre the section together to minimise the effects of shrinkage.

STEP 1 Look at the shape of the room and plan out which lengths to cut first. It is common practice to start with the longest walls; however, this is not always the most efficient method.

PRACTICAL TIP

Alcoves and bays are good starting points, as they are usually focal points in a room, and because the skirting is trapped between two walls. It is much easier to scribe out of a corner than to scribe both ends of a length of skirting.

Shrinkage on a scribe only shows in one direction so scribes facing into alcoves and bays will not be as obvious if the skirting shrinks.

Often the second piece of skirting to be fixed is also trapped between two walls. In this case scribe one end to sit over the first piece and continue around the room in the same way.

Check for water, gas and electrics before beginning fixing.

Check the floor for level: the skirting may need scribing level, always starting from the lowest point.

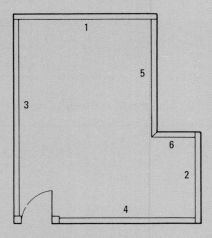

Figure 6.98 Order of fixing (trapped pieces first)

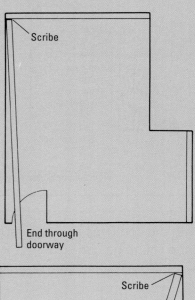

Figure 6.99 Extend through doorway to permit scribing of joint, or angle and scribe when both ends are trapped

STEP 2 Having decided on where to start, measure and cut the first pieces. These should be straight cuts with no joints if possible. Fix these lengths, unless they require scribing to the floor, in which case leave them loose at this stage.

STEP 3 Work out of the corners, scribing over skirting.

STEP 4 Mitre around external corners, checking the fit of each mitre before fixing.

PRACTICAL TIP

Try to avoid having lengths of skirting with scribes on both ends.

PRACTICAL TASK

19. MARK AND CUT AN EXTERNAL MITRE

OBJECTIVE

To fit and fix a mitre on a straight piece of skirting.

Skirting is mitred at external corners; a butt joint would expose end grain on one of the pieces which would look unsightly. Mitres are subject to double shrinkage, when both halves of the mitre shrink away from the corner.

STEP 1 Remove any build-up of plaster that may obstruct the joint or throw the skirting out of upright, as this will cause real difficulty when cutting joints and will look unsightly.

STEP 2 Place the piece of skirting in position and mark the top edge where it sits against the corner.

STEP 3 If you are using a purpose-made mitre box ensure you use the same saw you have made the original cut with when making your box. You can of course use a frame saw if the skirting is not too deep to enable this. Alternatively, use a chop saw, as long as you are trained and competent in its use.

Make the cut in a mitre box using a panel, frame or tenon saw, depending on the depth of cut. Alternatively mark and cut freehand or use a chop saw. Cut from the face so any breakout is on the back and will not show.

STEP 4 Repeat for the other half of the mitre.

STEP 5 Offer both pieces up to the corner and check the fit. Trim with a block plane if required until a good fit is achieved.

STEP 6 Complete any cuts that are required at the other end of the pieces and fix it back to the wall using the appropriate method for the background material.

STEP 7 Dovetail nail the mitre and punch all nails below the surface, adding a small amount of PVA adhesive to the join, if necessary.

20. CUT MITRING AROUND CORNERS THAT ARE UNDER OR OVER 90°

OBJECTIVE

To fit and fix a mitre on corner of skirting.

When cutting skirting around acute or obtuse corners, the angle can be determined by bisecting.

STEP 1 Place the skirting against the wall and mark the thickness on the floor for both halves of the joint. Where the lines meet, allow them to cross.

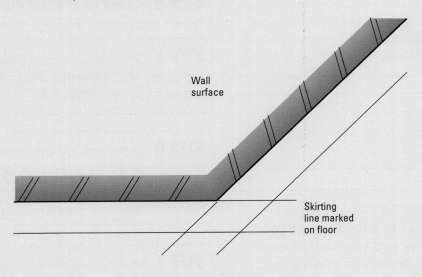

Wall surface

Skirting line marked on floor

Wall surface

Skirting line marked on floor

Figure 6.100 Line of skirting for corners over 90°

STEP 2 Place the skirting in position and mark the top edge, front outside edge and back inside edge.

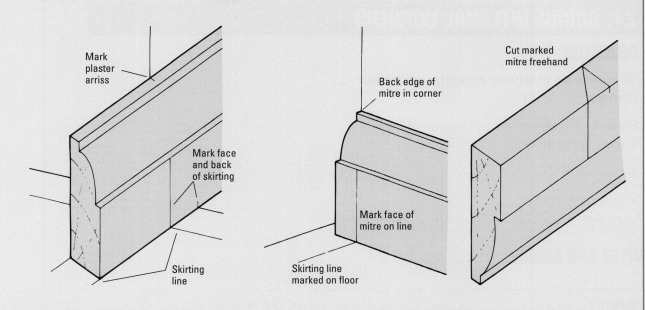

Mark plaster arriss

Mark face and back of skirting

Skirting line

Back edge of mitre in corner

Mark face of mitre on line

Skirting line marked on floor

Cut marked mitre freehand

Figure 6.101 Marking out external and internal corners over 90°

STEP 3 Mark a line across the front and back of the skirting with a try or combination square.

STEP 4 Set a sliding bevel to the required angle from the setting out on the floor and use this to set the required cut with your mitre box or chop saw.

STEP 5 Repeat the process for the other half of the joint.

STEP 6 Fit the joint with block plane if required.

STEP 7 Fix the skirting in position and nail the mitre as previously described.

21. SCRIBE INTERNAL CORNERS

OBJECTIVE

To learn two methods of scribing internal corners.

Scribing can be carried out by one of the following methods:

* mitre and scribe
* compass scribe

MITRE AND SCRIBE

STEP 1 Cut one piece of skirting into the corner. This should be a square cut but if the wall is not plumb the piece should be cut to fit.

STEP 2 Cut an internal mitre on the other piece. This will make the profile of the skirting apparent, outlining the shape of the cut required.

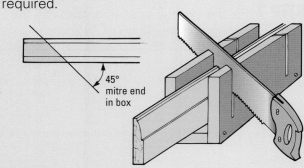

45° mitre end in box

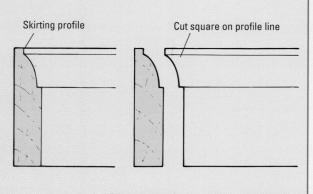

Skirting profile   Cut square on profile line

Figure 6.102 Cutting an internal scribe

STEP 3 Turn the skirting onto its top edge and cut the flat section of the profile with a tenon saw, slightly undercutting to ensure a tight fit.

STEP 4 Use a coping saw to follow the shape of the mould formed by the mitre on the top edge of the skirting, again slightly undercutting.

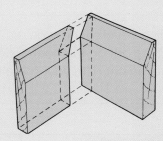

Figure 6.103 Scribing internal corners

STEP 5 Test the scribe against the skirting, making sure of a tight fit, and secure with appropriate fixings.

COMPASS SCRIBE

STEP 1 Cut one piece of skirting into the corner.

STEP 2 Offer the piece of skirting that is to be scribed against the first, set a compass and scribe to the face of the first piece of skirting.

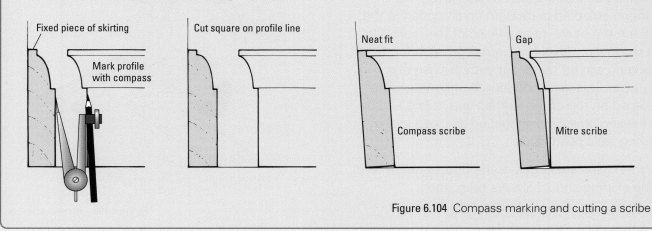

Fixed piece of skirting

Mark profile with compass

Cut square on profile line

Neat fit

Compass scribe

Gap

Mitre scribe

Figure 6.104 Compass marking and cutting a scribe

STEP 3 Cut to the shape of the scribe, following Steps 3 and 4 for the previous practical task.

Internal angles over 90° (obtuse)

The general rule with internal corners is to scribe due to single shrinkage; however, on obtuse angles the scribe becomes almost impossible to execute because of the clearance required behind the scribe. In these instances a bisected mitred joint should be used.

Heading joints

Jointing in length may be required when suitably long lengths of skirting are not available. Here you would use a heading joint. The skirting could be butted but this would result in an inferior finish and so it should always be mitred. The mitred heading joint makes any shrinkage in the skirting less obvious than would a butted joint.

The heading joint is simply two mitres placed on the ends of the skirting and jointed and fixed against the wall.

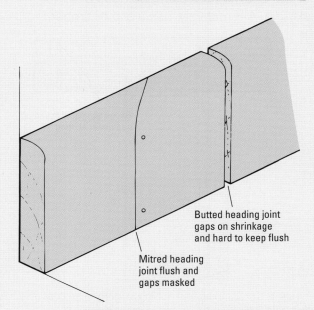

Butted heading joint gaps on shrinkage and hard to keep flush

Mitred heading joint flush and gaps masked

Figure 6.105 Mitres are preferred for heading joints

22. SCRIBING SKIRTING BOARD TO THE FLOOR

OBJECTIVE

To fit and fix a length of skirting to the floor.

Skirting boards should be fixed level; this is often neglected in modern construction to the detriment of the fit and finish of the work. On a reasonably level floor, small gaps under the skirting can be taken out with the use of a kneeler. This is a short length of board that is placed on top of the skirting and held down by kneeling on it. This should be as close to the fixing as is practicable.

When gaps are larger, or the floor is out of level, the skirting should always be scribed.

Figure 6.106 Using a kneeler

STEP 1 Holding the cut length of skirting in place, level and temporarily nail it in position, leaving the heads proud for removal.

STEP 2 Set a pair of compasses to the widest gap under the skirting and mark a line parallel to the uneven floor along the face of the skirting. Alternatively, use a slip of timber.

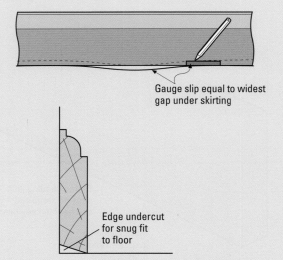

Figure 6.107 Scribing and cutting skirting to an uneven floor surface

STEP 3 Remove the skirting and using a hand saw or jig saw cut to the line, undercutting slightly for a tight fit on the front edge.

STEP 4 Fix the skirting in position.

23. INSTALL A PICTURE AND DADO RAIL

OBJECTIVE

To fix picture and dado rails to walls.

The methods for installing picture and dado rails are the same as for skirting; however, unlike skirting, a horizontal line is required around the room before installation can begin.

STEP 1 Mark a horizontal line around the room to a given height – this may be taken from the drawings, client instruction or governed by wall coverings such as wallpaper or panelling. This should represent the underside of the moulding.

If a datum is established on the site then the height will be directly related on the drawings.

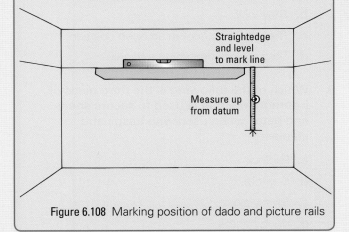

Straightedge and level to mark line

Measure up from datum

Figure 6.108 Marking position of dado and picture rails

STEP 2 When you are working alone, temporary nails should be knocked into the surface along the line to provide support for the moulding until it is fixed.

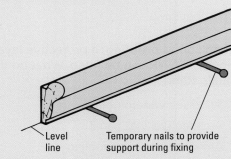

Level line

Temporary nails to provide support during fixing

Figure 6.109 Temporary support for dado and picture rails during fixing

STEP 3 Follow the procedure for fixing skirting.

STEP 4 Nail all external mitres.

TEST YOURSELF

1. What is another name for a portable powered mitre saw?

   a. Chop saw

   b. Hack saw

   c. Precision saw

   d. Timber saw

2. When sound-proofing, a protective layer, known as AWP, can be fitted. What do these initials stand for?

   a. Anodised waterproofing

   b. Architectural wall panel

   c. Architectural warp proofing

   d. Architectural waterproofing

3. What feature is added to access panels to protect them from damage each time they are removed or fixed back into position?

   a. Metal plate

   b. Plastic veneer cover

   c. Chamfered edges

   d. A thicker coat of paint

4. What material is commonly used to make a batten framework for a bath panel?

   a. Softwood

   b. Hardwood

   c. MDF

   d. Plastic struts

5. What can be used to ensure that base units are level and plumb when dealing with an uneven floor?

   a. Relaying the floor

   b. Fixing the base units more securely to the wall

   c. Adjusting the legs or wedges

   d. Fixing pegs to the next unit

6. Which is the most common type of kitchen worktop surface used?

   a. Post-formed

   b. Solid timber

   c. Marble

   d. Granite

7. Which of the following is likely to be a function of an external door?

   a. Security

   b. Weatherproofing

   c. Durability

   d. All of these

8. Which of the following is a true statement about a mortise latch?

   a. They are external latches

   b. They lock

   c. They hold the door in an open position

   d. They are opened using a handle

9. Which of the following is the third piece of ironmongery that is used to secure sheds and gates when used with hasps and staples?

   a. Padlock

   b. Keyhole

   c. Screw slot

   d. Knob handle

10. Which of the following timber mouldings is most likely to be closest to the cornice on a wall?

    a. Plinth block

    b. Skirting

    c. Picture rail

    d. Dado rail

Chapter 7

Unit CSA–L2Occ38
CARRY OUT STRUCTURAL CARCASSING OPERATIONS

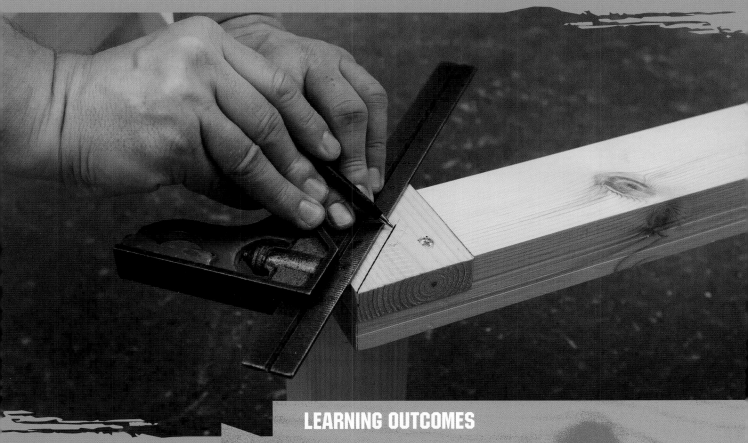

LEARNING OUTCOMES

LO1/2: Know how to and be able to prepare for structural carcassing operations

LO3/4: Know how to and be able to erect trussed rafter and traditional gable roofs

LO5/6: Know how to and be able to construct gables, verges and eaves

LO7/8: Know how to and be able to install floor joists

INTRODUCTION

The aims of this chapter are to:

* help you interpret relevant information

* help you to select resources to carry out the work

* show you how to erect structural carcassing in accordance with the work specification.

PREPARING FOR STRUCTURAL CARCASSING OPERATIONS

Structural carcassing is a term used to describe the carpentry required for the structural elements of a building. This includes the floors and roofs.

Many of the necessary preparations are similar to those required for first and second fix operations covered in Chapters 5 and 6. In each of the first parts of this chapter the information contained in the other chapters is highlighted. But there are some specific elements that need to be looked at in more detail.

Hazards

Some structural work will involve working at height. This means that falls from height and potential injuries from materials falling from height are a concern. Materials used are generally extremely heavy and will require you to be trained in manual or mechanical lifting operations, depending on the work.

It is the duty of the site manager to assess, eliminate and control the risks of anyone falling from height or lifting operations.

Structural carcassing will involve working on joists and roofs. The materials used are heavy and potentially fatal if they should fall on anyone. There is always the chance that the structure might collapse part-way through construction. This means:

* Any area on the site that is at risk from falling materials needs to be clearly marked. It must include an exclusion zone as well as other areas on site that are strictly hard hat areas.

* Any walkways need to be covered.

* Where possible high-reach machines should be used and the machine cabs need to be reinforced. Using these machines is a specialist job and the operative must be fully competent and have the correct qualification for the category of lifting equipment used. In some extreme cases a crane may be required.

Anyone working on the site needs to be trained, fully qualified and properly supervised. This means that a full risk assessment relative to each task needs to be carried out. Where appropriate further advice should be taken from the HSE.

Access equipment

It cannot be stressed enough how important it is to manage any work at height. The HSE has what it calls a hierarchy of controls: before taking responsibility to use any access equipment ensure you are both fully trained and competent in its use.

* Avoid – this means looking at other options, such as being able to do the work safely on the ground rather than at height.

* Prevent – if work at height is necessary, what can be done to make it less likely that someone will hurt themselves? This may mean putting in fall restraints and using harnesses.

* Arrest – if the worst should happen and someone does fall, restraints, fall bags and safety netting should be in place. These are arresting devices that break the fall.

There are various ways in which safe access to the roof area can be achieved. Table 7.1 outlines the main methods.

DID YOU KNOW?

Refresh your memory about preparing for work by referring to:
* Chapter 2 for drawings, specifications and schedules
* Chapters 1 and 5 for PPE and collective protective measures
* Chapter 5 for tools, equipment and materials
* Chapter 5 for timber defects.

DID YOU KNOW?

A third of all falls from height involve either ladders or step ladders. Ladders can be safe in some circumstances, but they are often misused.

Methods to access roofs	Description
General access scaffold	These are scaffolds that are independent of the structure. Correctly erected and regularly inspected, these can provide safe access, along with a working platform. They can also provide a place to store materials. They are erected by specialists and are constructed on firm, level ground. All scaffolds should be braced correctly to ensure the stability of the structure.
Stair tower and fixed or mobile scaffold towers	These, also known as tower scaffolds, are safer than ladders. They are usually made from aluminium components, although some are steel. The components are locked together to give considerable strength. They need to be properly secured and erected by a fully trained and competent operative.
Mobile access equipment	Many of these are referred to as mobile elevating work platforms (MEWP). A scissor lift will lift materials or objects vertically only. A telescopic boom is also known as a cherry picker, which lifts vertically and can also reach outwards. The final option is an articulating and telescopic boom, which is often mounted on a vehicle. These are all ideal for lifting objects while working at height. They should only be operated by trained and competent individuals who are qualified in the specific plant that is to be operated.

Methods to access roofs	Description
Ladders	Ladders can be used if, after assessing the risks, the use of more suitable work equipment is not justified because of low risk and short duration (up to about 30 minutes). When ladders are used they must always be of the right type (industrial grade) and there should always be three points of contact by the user: two feet and one hand on the ladder. They must be secure, tied at the top or bottom, in good condition and regularly inspected. The correct length of ladder should also be used.
Roof access hatches	These are essentially covers for openings for roof access ladders. They are designed to prevent people from falling through the holes. The best practice is to ensure that the roof hatches are always closed after use and only opened while workers or materials are being passed up or down through the hole.

Table 7.1

Protecting the work and its surrounding area

In Chapter 3, when we looked at sustainability, we learned that it is important to ensure that all construction work has the least possible negative effect on the surrounding area and environment. One of the ways this can be achieved is to keep the site clean and dispose of waste in an environmentally friendly way. This means recycling as much waste as possible.

ERECTING TRUSSED RAFTERS AND TRADITIONAL GABLE ROOFS

Truss rafter roofs

The majority of modern domestic dwellings have factory-made roof units. These are manufactured as triangulated frames, which are called trussed rafters. The timber is prepared and stress graded. Many of them have specific design names, such as 'fan' or 'fink'. Once these trusses have been positioned to the correct specification they need to be braced following the plan normally provided by the manufacturer.

Roofs are normally constructed using a number of trussed rafters. These are spaced at centres between 400 mm and 600 mm. Each of the roof trusses is designed to sit on a wall plate. The rafters are then fixed into place using truss clips. Further support is given by galvanised wall straps or restraints.

Typically, the erection of truss rafters follows a set procedure:

* The position of each of the truss rafters is marked on the wall plate.

* Each of the rafters is then lifted into place.

* The rafters are stacked upright at one end of the roof but not leaning against the gable-end as brickwork may not be strong enough to support it.

* The first rafter is then lifted into position and is secured with truss clips.

* The first rafter is plumbed and braced (temporarily) using diagonal braces and binders.

* Each of the remaining rafters is then put into position.

* Each of them is secured and temporarily braced.

* Once all of the rafters are in place the diagonal and longitudinal braces are then secured.

* Finally, any necessary straps and restraints are secured.

It is usually the case that gable end walls are either partially or fully completed before the truss rafter roof is erected. When this is the case a single trussed rafter frame or even a pair of common rafters are fixed and braced at each gable. These will act as a guide for the bricklayer to shape the top of the walls. If the building is to have gable ladders (which project over the face of the wall) the brickwork is only built up to the underside of the truss. The brickwork is then finished off once the gable ladder is in position.

Lengthening joints and half lap joints

Gable end roofs only need wall plates on the walls where the roof pitches. If the walls are greater than the standard timber length of 6.3 m, then it is necessary to join the wall plates with half lap joints. In these cases the lap will be equal to the width of the timber. In practice you should use 3 m lengths of timber to joint the wall plate with the half lap nailed together. These joints to the wall plate must be at least 100 mm in length.

Trussed rafters are frames made from stressed graded timber. They are held together with galvanised gang nail plates and engineered to carry their own proportion of the imposed load. Their main advantages are the clear span that can be achieved without the need for intermediate supporting walls, the speed of construction and the controllable manufacturing process. All of these factors add to the benefits above a traditional cut roof.

The exact size and spacing of rafters could be determined by the imposed load on the structure. In some areas regular heavy snowfalls or strong winds will mean that the roof may have to cope with heavier loads. This will mean that the rafters have to be spaced closer together, have shorter spans and are larger in section. These factors are generally determined by the manufacturer after the order is placed.

Traditional roofing

There are various ways to determine the actual length and bevels for the roof. For small gable end roofs a full size setting out can be used or the setting out can be made to a suitable scale. This involves working on a level surface, as can be seen in Fig 7.1.

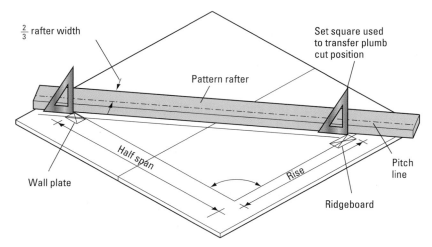

Figure 7.1 Marking pattern rafter to scale

As can be seen in the diagram, provided the measurements are carefully checked it is possible to show the rise and run of the rafter and the diagonal or pitch line.

Erecting the gable roof

The steps in erecting a gable roof can be seen in Fig 7.2.

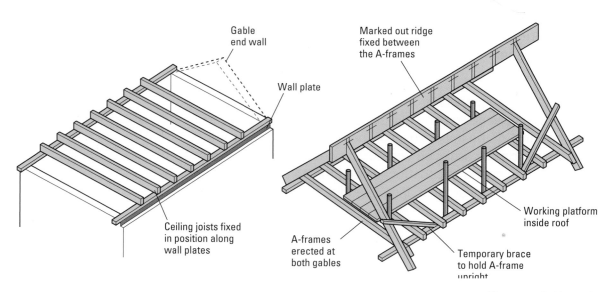

Figure 7.2 Gable roof erection

Fig 7.3 outlines the span of pitched roof rafters and ceiling joists.

Based on using C16 grade softwood rafters for roofs over 30° and up to 45° pitch, with a dead load of up to 0.5 kN/m², imposed loading of 0.75 kN/m² suitable for most sites located up to 100 m above ordnance datum with access only for maintenance and repair. Ceiling joists are for dead loads between 0.25 and 0.50 kN/m².

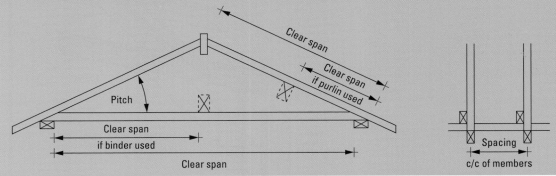

Size of rafter (mm) B × D	Spacing (centre to centre)		
	400 mm	450 mm	600 mm
38 × 100	2.28	2.23	2.10
38 × 125	3.07	2.95	2.69
38 × 150	3.67	3.53	3.22
50 × 100	2.69	2.59	2.36
50 × 125	3.35	3.23	2.94
50 × 150	4.00	3.86	3.52
Size of ceiling joist (mm)			
38 × 72	1.11	1.10	1.06
38 × 97	1.67	1.64	1.58
38 × 122	2.25	2.21	2.11
38 × 147	2.85	2.80	2.66
50 × 72	1.27	1.25	1.21
50 × 97	1.89	1.86	1.78
50 × 122	2.53	2.49	2.37
50 × 147	3.19	3.13	2.97

Figure 7.3 Maximum clear span of pitched roof rafters and ceiling joists

Wall plates

Wall plates are usually secured to the wall with metal restraint straps. The rafters are nailed to the wall plates and further lateral stability is provided with restraining straps that are secured to the rafters and the gable wall blockwork.

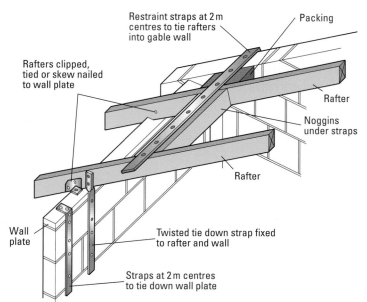

Figure 7.4 Anchoring of rafters and wall plates

Labels on figure:
Restraint straps at 2 m centres to tie rafters into gable wall
Packing
Rafters clipped, tied or skew nailed to wall plate
Rafter
Noggins under straps
Rafter
Wall plate
Twisted tie down strap fixed to rafter and wall
Straps at 2 m centres to tie down wall plate

Lateral restraints

In cases where gable ladders are not used, restraint straps at both rafter and ceiling level are used to prevent movement and cracking of walls. The straps have a minimum cross-section of 30 × 5 mm. They are fixed under the rafters and over the ceiling joists. Noggins are installed under the straps and the straps are fixed to a minimum of three trusses. Each of the straps is fixed in place with nails or screws or built into the gable wall later, if not yet constructed..

Timber bracing

As can be seen in Fig 7.5, which shows the erection procedure for trussed rafters, bracing can be incorporated at various points to support the roof.

There are various bracing requirements, which are specified in the design drawings across the trussed rafter roof.

* Temporary longitudinal bracing – 75 × 25 mm or 100 × 25 mm – can be used to stabilise the trusses while they are being erected.

* Permanent diagonal bracing should be installed, creating a 45° angle to the rafters. They should run from the highest point on the underside of a truss to overlap and fix to the wall plate. This needs to start at the gable end and zigzag down the length of the roof. It is fixed to every truss. It is recommended that there should be at least two of these braces on each slope. They are nailed into position with annular ring shank nails.

* Longitudinal bracing is positioned at **node points** along the length of the roof.

* Diagonal chevron bracing is used for larger roofs and should cover at least three trusses.

The practical task describes how to erect trussed rafter roofs.

PRACTICAL TASK

1. ERECT TRUSSED RAFTER ROOFS

OBJECTIVE

To erect a series of trussed rafters to form a roof.

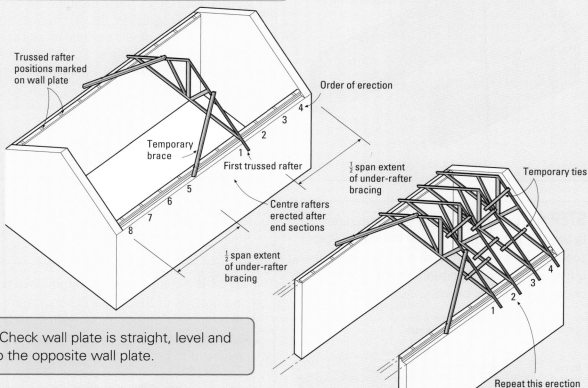

Trussed rafter positions marked on wall plate

Temporary brace

First trussed rafter

Centre rafters erected after end sections

½ span extent of under-rafter bracing

Order of erection

½ span extent of under-rafter bracing

Temporary ties

Repeat this erection sequence on other end of roof

Figure 7.5 Trussed rafter roofs erection procedure

STEP 1 Check wall plate is straight, level and parallel to the opposite wall plate.

STEP 2 Mark out one wall plate. The first truss should be set 50mm from the inside of the internal blockwork and the second at 400 to 600mm from the inside face of the blockwork. All other trusses are fixed at 400 to 600mm centres. Then maintain centres from the second truss.

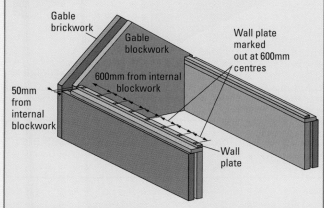

Gable brickwork

Gable blockwork

Wall plate marked out at 600mm centres

600mm from internal blockwork

50mm from internal blockwork

Wall plate

Figure 7.6 Marking out the truss positions along the wallplate

TOOLS AND EQUIPMENT

Tape	Hand saw (cross cut)
Pencil	Spirit level, straight
Framing square	edge and or plumb bob
Combination square	Power tools
Claw hammer	Nail gun

PPE

In this task, and the ones that follow, ensure you select PPE appropriate to the job and site where you are working. Refer to the PPE section in Chapter 1.

STEP 3 Transfer marks from the wall plate onto a longitudinal brace. Use the brace as a template to mark out the opposite wall plate. Ensure the template is the right way around.

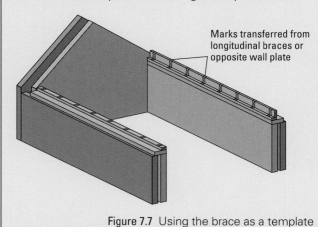

Marks transferred from longitudinal braces or opposite wall plate

Figure 7.7 Using the brace as a template

STEP 4 Stack the number of longitudinal braces required to complete the work. Place the template on the top of the stack and, using a framing or roofing square, transfer the marks on to all the other braces.

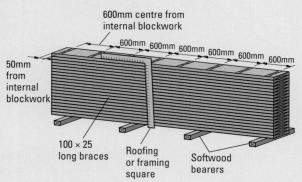

600mm centre from internal blockwork

600mm 600mm 600mm 600mm 600mm 600mm

50mm from internal blockwork

100 × 25 long braces

Roofing or framing square

Softwood bearers

Figure 7.8 Using pattern brace to mark out longitudinal braces

STEP 5 Place truss clips on top of the brickwork next to each mark on the wall plate, on both sides of the roof.

STEP 6 Place the first trussed rafter, and fix it at the wall plate using truss clips (nail all holes). Plumb the truss and temporary brace back to the wall plate on both sides of the roof.

STEP 7 Place the second truss, and fix it with truss clips.

PRACTICAL TIP

Where trusses are very large a crane will be needed to lift them into position one at a time as the work progresses. Smaller trusses that can be picked up, should be lifted on and stacked vertically at the gable end of the roof, if possible, but not leaning against a modern cavity wall end.

PRACTICAL TIP

By marking out all the longitudinal braces in advance, measuring is eliminated when fixing, leaving both hands free to fix.

STEP 8 Fix a temporary longitudinal brace on the outside of the first and second trusses and allow it to sit on the gable wall. If each truss is being lifted in with a crane, temporary braces will have to be kept short to allow access.

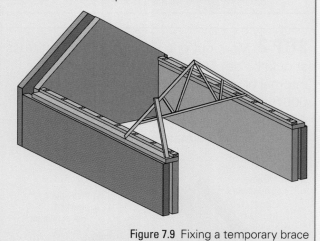

Figure 7.9 Fixing a temporary brace

STEP 9 Continue as above until all trusses are positioned.

STEP 10 Fix the permanent diagonal braces. The top of the brace should be as high up in the first rafter as possible, and it should also sit over the wall plate close to the gable end at the bottom. Fix braces to the underside of all intermediate trusses. Repeat on the other side of the roof.

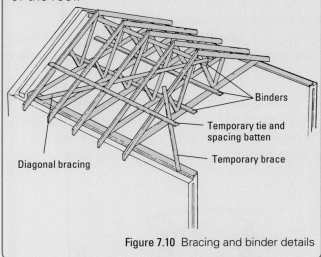

Binders

Temporary tie and spacing batten

Diagonal bracing

Temporary brace

Figure 7.10 Bracing and binder details

STEP 11 Fix longitudinal braces to the tops of the ceiling ties as specified on the drawing and in the schedule. Make sure they are fixed on the centres previously marked in Step 4.

STEP 12 Fix all remaining longitudinal braces. The ridge brace should be fixed as high as is practicable.

Longitudinal braces are fixed adjacent to node points.

STEP 13 The process is now repeated along the rest of the roof. If additional trusses are required these can be braced off the completed section.

STEP 14 Fix restraint straps to the rafters and ceiling ties. These should have a minimum of four fixings and at least one should be in the third rafter/joist.

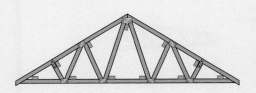

NB other bracing omitted for reasons of clarity

Longitudinal brace located at all nodes (excluding support) and nailed to every truss using 2 No. 3.35 mm × 65 mm galvanised wire nails.

Figure 7.11 Longitudinal bracing

Transporting and lifting truss rafters

Trusses can easily be damaged by poor handling. They should be inverted when they are being transported. Generally it is recommended that the trusses are lifted either at the rafter (top chord) or ceiling tie (bottom chord). These are called are node points. They are where the joints are made. It is not usually recommended to lift individual trusses.

When they are being moved around on site, particularly by machinery, they need to be banded together and supported using a **spreader bar**.

When the trusses are being handled on site no fewer than three people should assist. Trusses can weigh a considerable amount, even though it may not look as if they are heavy:

Trusses need to be carefully stored on site in order to prevent them from being deformed or damaged. They can be held in position over a **trestle prop** for vertical storage. Although they can be stored horizontally, problems may occur when lifting up to the vertical because the truss can bend as it is angled and the gang plates may be loosened on the joints.

Fig 7.12 shows ways of handling trussed rafters. Ensure there are enough colleagues to move them without damage or injury. They can be lifted into position by crane controlled with a guide rope from the ground, or the roof can even be pre-assembled at ground level.

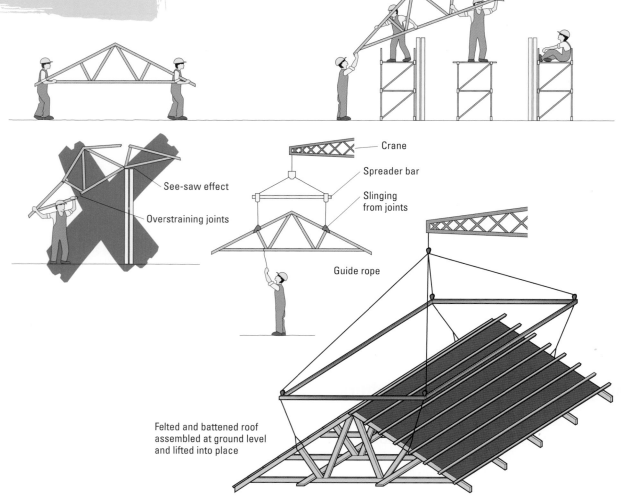

Figure 7.12 Handling trussed rafters

Openings in truss rafter roofs

There will almost certainly be a requirement to provide openings in a roof. This can be for a variety of different purposes:

* roof lights or dormer windows

* chimneys

* loft access hatches.

In these cases either the rafters or the ceiling joists will have to be trimmed in order to provide the opening. These are designed into the roof at the drawing stage if the openings are being made retrospectively. Calculations need to be produced by a construction engineer. Rafters, joists and trimmers are held together with joist hangers and housing joints. In the case of chimneys, it is also important to ensure that there is a gap between the brickwork and any structural timbers. In practice this is 40 mm. The purpose of the gap is not only to provide ventilation, but also to reduce the exposure of the timber to hot brickwork in the chimney, preventing fire. The area around the chimney stack on the roof has to be waterproofed and this means fixing a framework to support any gutter or flashings, as can be seen in Fig 7.13, which shows how this trimming is achieved.

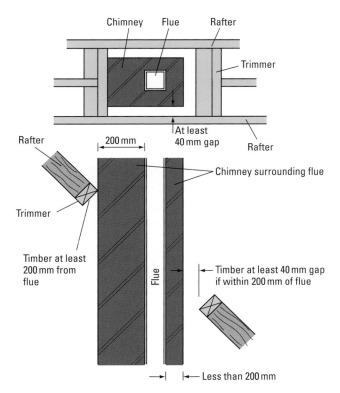

Figure 7.13 Separation of flue and combustible materials

Erecting truss rafters at ground level

In order to speed up the building of dwellings, trussed rafters can be constructed at ground level and then craned into position (see Fig 7.12 on page 252).

(see Fig 7.12 on page 252).

PRACTICAL TASK

2. TANK STANDS

OBJECTIVE

To construct water tank support for standard fink trusses.

Refer to previous practical tasks for tools and equipment.

PPE

In this task, and the ones that follow, ensure you select PPE appropriate to the job and site where you are working. Refer to the PPE section in Chapter 1.

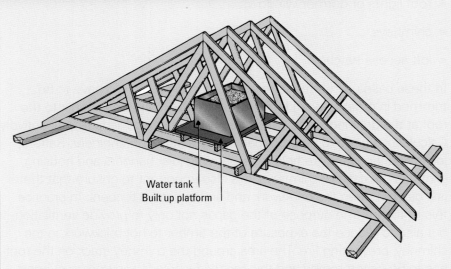

Water tank
Built up platform

Figure 7.14 Water tank platform

STEP 1 Fix two bearers as close to the node points as possible across ceiling ties. Consult the drawings and schedule for specified sizes and bearing.

STEP 3 Cut a further two bearers to sit at 90° to the second pair of bearers. These should be fixed at the width of the tank they are to support.

STEP 2 Cut two further bearers to sit at right angles to the first bearers. These should be spaced to the length of the tank.

STEP 4 Close board the bearers with 100 × 25 mm softwood boards or a 25 mm plywood sheet cut to size. Fix with annular ring shank nails or screws.

Raised water tank stands

For raised water tank stands, the first pair of bearers is substituted with frames made to the required height. These frames should be cross braced and of sufficient sectional size to support the tank and its contents. Consult drawings and schedules to determine the size of the timbers and the bracing conditions.

GABLES, VERGES AND EAVES

Gables, verges and eaves are all features of a pitched roof, along its edge. These are ways in which the overhanging part of a gable end roof can be finished.

Often the roof will overhang the wall of the building. This is known as the verge. An additional rafter and noggins extend beyond the gable wall. The term gable ladder refers to the frame assembly, which can be seen in Fig 7.15.

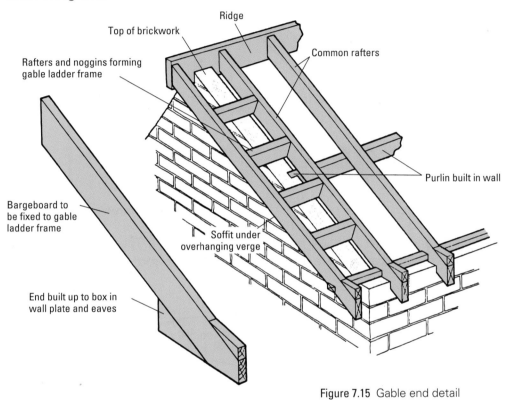

Figure 7.15 Gable end detail

The traditional way of constructing a gable ladder is to fix it on site by having it built in by the bricklayer and lining it with a soffit board on the underside. It would have bargeboards on the face side.

These days, gable ladders are often supplied prefabricated and are delivered to site along with the trussed rafters. If gable ladders are not supplied they will have to be manufactured on site as described previously. The method of positioning and fixing is the same for both. When the ladders are made on site accurate measurements are essential, referring to the drawings, specification and schedule.

Bargeboards

A bargeboard is basically a continuation of a fascia board. This is fixed around the verge, or the sloping edge of the roof. The lower end of the bargeboard is designed to box in the wall plate and eaves. This process needs careful measurement.

One way is to mark the bevels at the apex and foot of the bargeboard after temporarily fixing the board into position. As can be seen in Fig 7.16, a spirit level can be used.

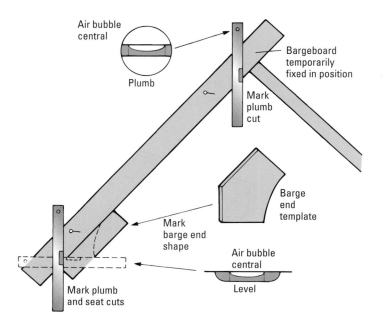

Figure 7.16 Marking out a bargeboard

The alternative is to use adjustable bevel squares. These can be set at the required angle in relation to the pitch of the roof, as can be seen in Fig 7.17.

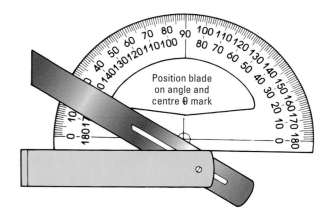

Figure 7.17 Setting an adjustable bevel to a known angle

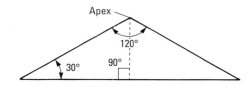

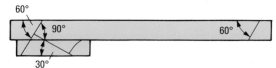

Figure 7.18 Determining angles for a bargeboard

> **PRACTICAL TIP**
>
> Remember that the total of all three angles in any triangle is always 180°. In the diagram the pitch is 30°. To find the apex angle: 30° + 90° = 120°; 180° – 120° = 60°. Each bargeboard needs to be half the angle of the apex. So this means that they need to be 60° each.

Bargeboards tend to be either butted with an overhang, or mitred to the fascia board, as can be seen in Fig 7.19.

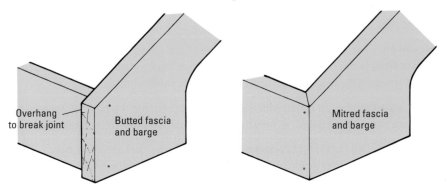

Figure 7.19 Jointing barge to fascia board

The most professional finish is the mitred joint. This is achieved by temporarily fixing the fascia and bargeboard into position. A length of timber that is the same thickness as the fascia and bargeboard is used to mark a pair of lines across the edge of the board. Obviously the angle will be 90°. Fig 7.20 and Fig 7.21 show how this is achieved and how bargeboards can be lengthened by adding additional timbers if sufficiently long lengths are not available.

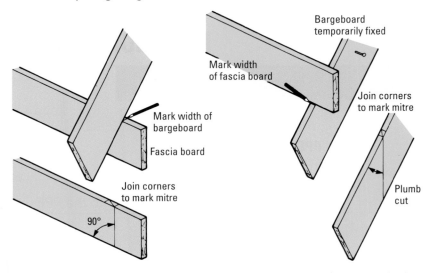

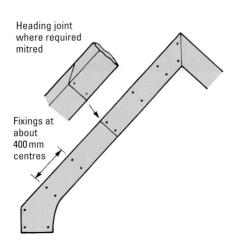

Figure 7.20 Making barge to fascia board mitre

Figure 7.21 Bargeboard lengthening/fixings

REED TIP

Learning skills is a lifelong experience; we're all still learning, and even after your apprenticeship is finished, you'll keep on learning too.

Finishing eaves

Eaves are where the rafters end on the lowest part of the roof slope. They will overhang the wall to some extent and need to be finished off with a fascia board and soffit.

Table 7.2 explains the purpose and fixing of five different ways in which eaves can be finished.

Eaves finish	Description and fitting
Flush	The rafters are cut off between 10 to 15 mm beyond the brickwork face. Fascia board is then nailed to them and this forms the support for the gutter. There will be a ventilation gap between the fascia board and the brickwork.
Open	To give the building weather protection the rafters extend well beyond the brickwork face. The rafters are in full view from the ground. In some cases the rafter then receives nailed-on fascia boards, but in other cases gutter brackets are fixed directly onto the rafters.
Closed	Closed eaves are very similar to open eaves except that the underside of the eaves is closed off with a soffit. A continuous vent strip is fitted to the back of the fascia. This carries the front edge of the soffit. The back sits on the brickwork and is wedged using noggins fixed between the rafter ends.
Sprocket piece	Sprocket pieces are nailed onto the top of each rafter, with the purpose of reducing the pitch of the eaves. The idea is to slow down the rate at which rainwater enters the gutter in extreme weather conditions.
Sprocketed	One of the problems with a steep pitched roof is that rainwater comes down the roof and can often overshoot the gutter. Sprockets are nailed to the side of each rafter in order to slow down the rate of flow. This means that the water, when it reaches the eaves, is more likely to go into the gutter.

Table 7.2

Joints, verges and eaves

The following practical task looks at how to fix gable ladders and bargeboards, working to specific drawings and customer requirements.

PRACTICAL TASK

3. FIX GABLE LADDERS

OBJECTIVE

To fix gable ladders and bargeboards.

See previous tasks for details of the tools and equipment that you will need to complete this exercise.

PPE

In this task, and the ones that follow, ensure you select PPE appropriate to the job and site where you are working. Refer to the PPE section in Chapter 1.

STEP 1 On prefabricated gable ladders the (vertical) plumb cut at the ridge intersection will be pre-cut. However, if the ladders are made on site, mark the plumb cut on both rafter ends and cut.

STEP 2 Position and fix the gable ladders in position. They can be nailed or screwed, or for very heavy or wide ladders they can be bolted to the last truss.

STEP 3 Repeat for the remaining ladders.

STEP 4 Temporary support may be required to prevent the ladder from sagging while the brickwork is completed.

PRACTICAL TASK

4. VERGE AND EAVES FINISHES

BARGEBOARDS

A bargeboard is very similar to a fascia board. It is fixed to the same angle as the roof and provides a decorative finish to the gable end.

There are two methods of determining bevels and lengths:

1. Marking in situ: the barge board is temporarily fixed in position and a spirit level is used to mark the (vertical) plumb and (horizontal) seat cuts in their respective positions.

2. Setting the bargeboard angles from the drawings.

METHOD 1: MARKING IN SITU

STEP 1 Temporarily fix the board in position.

STEP 2 Using a spirit level mark the plumb cut at the ridge and the seat cut at the intersection of bargeboard to fascia.

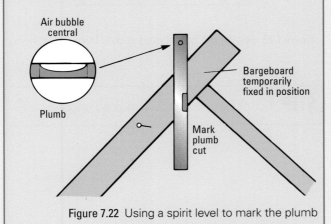

Air bubble central

Plumb

Bargeboard temporarily fixed in position

Mark plumb cut

Figure 7.22 Using a spirit level to mark the plumb

STEP 3 There are three methods of jointing the bargeboard to the fascia board: mitred, butted flush or butted and extended past the fascia.

Mitred methods 1 The angle is determined by temporarily fixing the pieces in position in turn and using a short end of timber the same thickness as the bargeboard and fascia board to mark two lines across the edge of the board. Join the opposite corners to form the mitre. The fascia board will have a face cut of 90° and the bargeboard will have a face cut that is plumb (vertical). Mark as on Fig 7.20.

Butted methods 2 and 3 Mark the face of the fascia board on to the face of the bargeboard; either cut flush or add an allowance for overhang. See Fig 7.19 for a diagram of bargeboards butted with an overhang, and mitred to the fascia board.

STEP 4 Cut a plumb cut at the top of the bargeboard.

STEP 5 Fix the bargeboard at 400 mm centres, double nailing using oval or lost head nails. These should be at least 2½ times the thickness of the bargeboard. Punch all nails below the surface.

STEP 6 When the bargeboard is to be fitted against boxed or closed eaves, it will require a shaped piece to be fitted to the underside of the bargeboard at the eaves to close off the boxed eaves.

STEP 7 Where the ladder frame extends past the gable end brickwork the underside of the ladder will need closing off. This is the verge. The verge will be finished in the same materials as the soffit.

METHOD 2: SETTING THE BARGEBOARD ANGLES FROM THE DRAWINGS

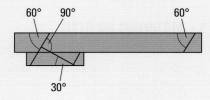

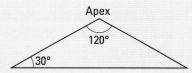

Figure 7.23 Setting the bargeboard angles from the drawings

STEP 1 Set bevels to the required angles.

STEP 2 Cut a plumb cut for the joint at the ridge or apex.

STEP 4 Mark out the seat and plumb cut at eaves.

STEP 3 Measure the length of the bargeboard from apex to fascia board.

STEP 5 Cut to length and fix as above.

Jointing in length (heading joint)

Where the bargeboard exceeds stock lengths of timber of the required section, it will need to be jointed in length. This is called a heading joint. Simply mitre the two lengths end to end as in skirting boards.

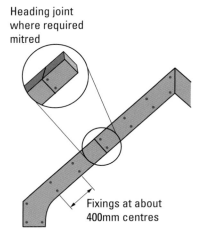

Figure 7.24 Heading joint and fixing suggestion

CASE STUDY

Learning as you work

Joshua Richardson is an apprentice in his third year at Laing O'Rourke.

'I haven't had the chance to work on roofs yet, but I'll be sent out for a special placement for a few weeks so I can get some real experience. I've worked on floor joists at the site I'm on though. It's quite repetitive and you can get into a rhythm – once you've done one section you follow that lead all the way through.

Always watch the guy who's teaching you – watch everything he does because sooner or later he's going to ask you to do it on your own. If you don't watch, and he goes off for say 20 minutes and comes back and you've not done it… Well, you're learning, but if your supervisor asks you if you know what you're doing, make sure you do!

If you make a mistake, and you can mend it, then it's not really a mistake. It's just something you learn from. And I'd rather get it right, it's quality that matters at this stage of your career, not quantity.'

FLOOR JOISTS

Joists are timber or purpose made beams arranged parallel to one another. Their purpose is to carry the floor and ceiling finish and to be able to carry imposed loads such as universally distributed loads (UDL) and point loads. They will also be required to carry staircases for upper floors. The Building Regulations state that all floor joist trimmers and trimming joists should be of a specified grade. This will be indicated in architects' drawings.

Types of floor joists

The diagram shows a typical cross-section of upper-floor construction.

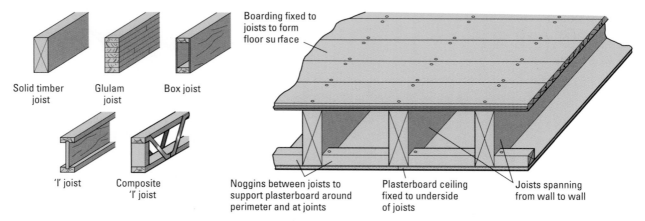

Solid timber joist

Glulam joist

Box joist

'I' joist

Composite 'I' joist

Boarding fixed to joists to form floor surface

Noggins between joists to support plasterboard around perimeter and at joints

Plasterboard ceiling fixed to underside of joists

Joists spanning from wall to wall

Figure 7.25 Joists

I-beam (designed floors)

A designed floor is engineered. The engineer supplies the I-beam manufacturer with details such as spans and openings to stairwells. The designer calculates the desired section sizes from this information. An I-beam can actually come in a variety of different materials. They are very efficient and have a very good strength to weight ratio and are more resistant to movement than traditional timber joists.

They have a far superior strength to weight ratio than many traditional joists and they use a lot less material.

They are lightweight, strong and available in long lengths and so are considered to be very quick and easy to install. They are delivered on site as a precise component, which reduces the installation time. I-beams usually have service holes already formed, with centres that can be knocked out with a hammer. The tradesperson should not drill or cut the I-beams in any way not specified by the designer.

Solid timber

When using solid timber it is important that structurally graded material is used. On many construction sites solid timber joists are being replaced with I-beams. This is due to the fact that solid timber joists have the disadvantage of splitting, warping and twisting.

However, solid timber joists have been used for centuries. The lack of availability of large enough trees often means that the span that they can bridge is also limited. Despite this, solid timber joists can still be effective if the span is increased with the use of supporting wall binders.

Laminates

LVL, or laminated veneer lumber, is an engineered timber product. Thin laminates, or wafers, of timber are glued together and then compressed. This creates a very high strength beam. The material is very resistant to bending, tension and compression. They are also able to carry heavy loads. They can be used for a variety of purposes and are a very real alternative to steel beams. They are much stronger than structural timber.

Glue laminated timber, or glulam, is increasingly being used. Sometimes it is referred to as laminated stock. It comprises several layers of timber that have been bonded together with a strong adhesive that is moisture resistant.

Supporting joists

There are a number of different ways in which the ends of joists can be supported. Each has a specific use and a specific reason why it might be used.

Built-in support

The minimum bearing in a wall should be 90 mm. The bearing refers to the amount of joist that sits in the wall. It allows for any defection in the joist which could collapse if the joist slips off its bearing.

When the joist end is being put into lightweight blocks, a steel bearing bar is put into the mortar joint.

Joist hangers

This is the preferred technique for most modern housing. The joists are supported by galvanised steel joist hangers.

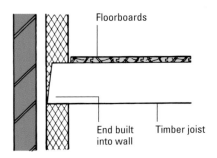

Figure 7.26 Building in a joist

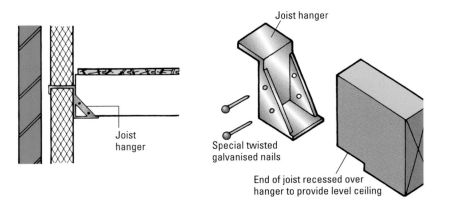

Figure 7.27 Use of a joist hanger

PRACTICAL TIP

It is a necessary precaution to make sure that the joist end does not project into the cavity. By creating a bridge across the cavity, moisture can, through capillary action, pass across the cavity. A telltale sign of this is the pattern staining on the internal plaster finish.

The joist itself is either built into the wall or it bears on a wall. Joist hangers can be walled into the brick or blockwork courses at appropriate centres.

Wall plates

Wall plates that support joists in ground floor construction can be found on the dwarf walls and the intermediate honeycomb sleeper walls and around fenders on abutments to fireplaces. Ceiling joist and attic floor joists are supported on the eaves wall plate. The wall plate distributes the load and provides a means of fixing.

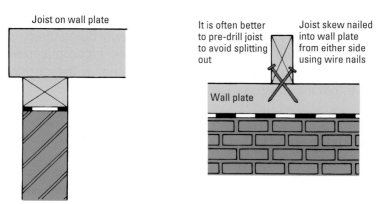

Figure 7.28 Joist supported on a wall plate

Normally a pair of joists meets over a wall plate. These are then nailed together so that they overlap the wall plate. The joists are secured in place using round-headed wire nails.

Binders

In large span construction of floors it may be necessary to incorporate intermediate support in the form of binders. The binders can be made of a variety of different materials:

* concrete

* steel universal beams (UB)

* timber, which can either be solid, glue laminated or oriented strand board (OSB) plywood.

The binders are either put in position beneath the joists or at the same level as the joists, as can be seen in Fig 7.29.

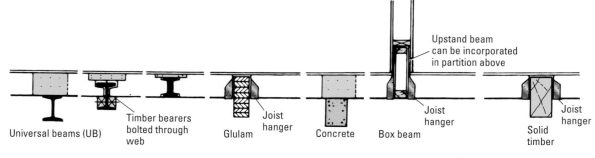

Figure 7.29 Various binders

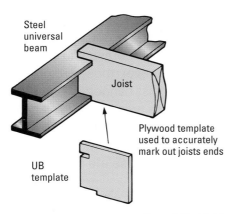

Steel
universal
beam

Joist

Plywood template
used to accurately
mark out joists ends

UB
template

Figure 7.30 Making a joist to fit a beam

Forming openings

Wherever there are openings the joists will need to be framed or
trimmed around them, as can be seen in the diagram.

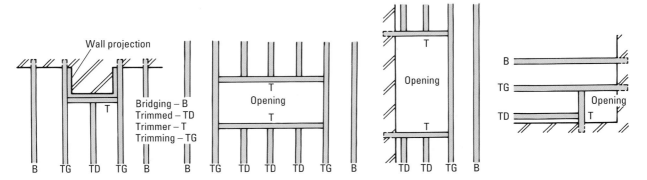

Figure 7.31 Trimming openings

Each of these different trimming techniques have a specific function, as
can be seen in Table 7.3.

Trimming technique	Function and description
Bridging joist	This is a joist that spans from one support to another. It is more generally known as a common joist.
Trimmed joist	This is trimming joist that has been shortened to form an opening and is supported by the trimmer.
Trimmer joist	This is a joist that is at right angles to a bridging joist. It is there to support the shortened ends of the trimmed joists.
Trimming joist	This joist supports the ends of trimmer joists and has the same span as a bridging joist.

Table 7.3

Joints between trimmer and trimming joists are now largely formed using heavy-duty joist hangers. The traditional methods of jointing these components, using tusk, mortise and tenon joints, can be seen in the following diagram. However, these are now only really seen in heritage work. By contrast, joist hangers are thin, galvanised steel and they are nailed into place with twisted nails, as can be seen in Fig 7.33. These joints are simply wedged into place. The joints between trimmed joists and the trimmer are housing joints but again these are rarely used in modern construction. These are fixed with wire nails.

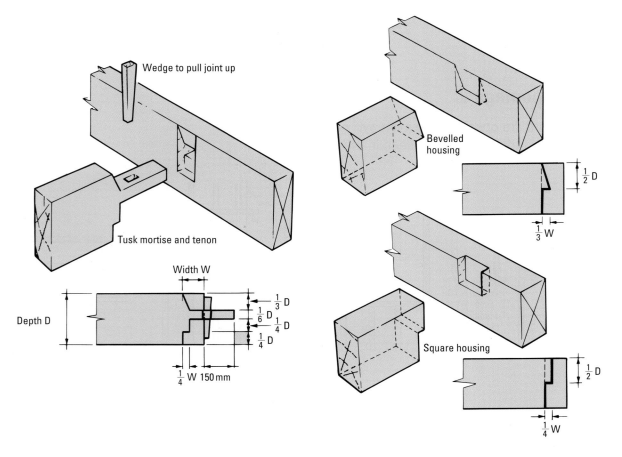

Figure 7.32 Traditional trimming joints

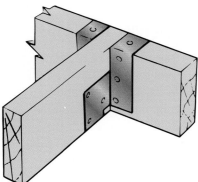

Figure 7.33 Trimming using a joist hanger

Strutting

Building Regulations require joists to be protected by struts. Joists that are between 2.5 and 4.5 m will need a strut at the midpoint. For joists in excess of 4.5 m two sets of struts are installed. One is placed at a third of the span and the other at two-thirds.

There are several different types of strutting that can be used, as can be seen in Fig 7.34 and in Table 7.4.

Type of strutting	Description and use	
Solid	This should be at least 38 mm thick. The strut needs to extend to at least 75 per cent of the depth of the joist. Solid strutting is relatively fast to install. However, it does loosen if the joists shrink. This is an inferior technique to herringbone strutting.	Section Plan — Solid timber strutting Struts staggered either side of centre line to permit easy nailing
Herringbone	The recommended minimum section size is 38 × 38 mm. They take longer to install but are still effective when the joists shrink. They can only be used when the space between the joists is less than three times the depth of the joist.	Nailed at centre Section — Nailed to joist Plan — Herringbone timber strutting
Galvanised steel	This is used in herringbone strutting and the galvanised steel can be ordered in ready-cut to size or can be purchased as adjustable to different lengths of joist.	Nailed to joist Section Plan — Galvanised steel strutting

Table 7.4

Galvanised steel herringbone strutting

Steel strutting is much easier to install than timber herringbone and solid strutting. It comes in lengths that can be bent on indentations in the steel for different joist spacings. Steel herringbone strutting, like its timber counterpart, allows services to be routed easily.

Perimeter noggins

Noggins should be cut and fixed between joists to carry plasterboards. In addition it is good practice to fix noggins at plasterboard centres along the joist length.

Restraining straps

Restraining straps should be fixed across at least three joists and should be notched and fixed to the tops of the joists. They can be built in or fixed to the face of the blockwork.

Joist positions

When the carpenter arrives to begin installing joists on a new build they usually continue a construction process that has already been begun by the bricklayers. Bricklayers and carpenters often work closely together to make progress run smoother. Conflicts and remedial work can be avoided if the process is shared and discussed in advance. It is in the best interests of the carpenter to ensure that the wall plates or joist hangers are level and spaced correctly.

PRACTICAL TIP

It is rare for the joists to work out exactly on centre. The last spacing can be under the prescribed centres but not more.

* The brickwork will be in place up to the underside of the built-in floor joist or bedded wall plate.

* Joist hangers will have already been fitted into the brickwork at the required height.

* If lugged hangers are being used as opposed to restraint hangers, the wall may already have been built up. The joint, ready to take the hanger, may have already been raked out.

The various options open to the carpenter can be seen in Fig 7.34.

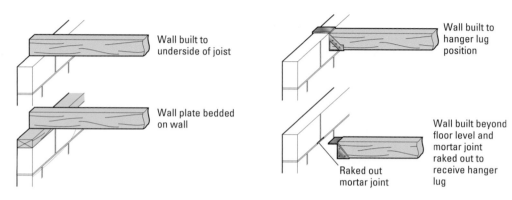

Wall built to underside of joist

Wall plate bedded on wall

Wall built to hanger lug position

Wall built beyond floor level and mortar joint raked out to receive hanger lug

Raked out mortar joint

Figure 7.34 Vertical position of joists

After measurement the joists are cut to length:

* A 20mm gap at both ends is necessary for ground floor joists.

* Splayed ends are advisable for upper floor joists to prevent the joist protruding into the cavity.

* Joists going into hangers have to be the exact length otherwise the hangers will not be tight against the wall. However, the joists should not be cut too tight as this will push out freshly laid blockwork, termed in the industry as 'green'.

* A 50mm gap is left between the outside joist and the wall. The outside joist should be the first one to go in. The first joist is set 50mm from the wall to allow for airflow. The second joist is set on centre from the wall. Subsequent joists are laid on centre from the second joist.

* All other joists are then laid out to centres at 400–600mm, depending on the load. The joist at the opposite wall should be 50mm from the wall.

The different options in cutting joists to length and things to avoid can be seen in Fig 7.35.

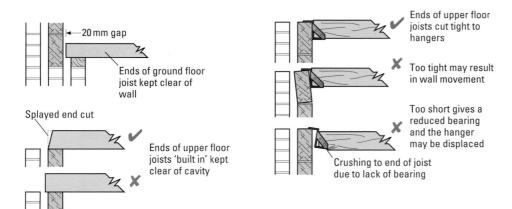

Figure 7.35 Cutting joists to length

Protection from moisture

Timbers that are used for structural work need to be treated with a preservative. The preservatives prevent fungi from forming. This in turn prevents rot.

Suspended timber flooring

Suspended timber flooring is the term used to denote the construction of ground and first floors in domestic and older commercial and industrial buildings. Informal terms include hollow and upper floors.

Classification of timber floors

Timber floors can be classified in three ways:

Single floors: only one type of joist is used – bridging or common joists.

Double floors: used where the shortest span in the floor is over 4.8 m. The bridging or common joists are supported by a large binder to give strength.

Triple floors: these are also called framed floors. Greater spans can be achieved using this method; however, they are rarely seen in modern construction. The bridging or common joists are supported by binders, which in turn transmit their load to a girder joist, steel or concrete beam.

The single ground floor

Ground floors usually consist of bridging joists which span between supporting walls. They are often supported on intermediate supporting walls called honeycomb sleeper walls.

Some trimming may be required around hearths using trimming, trimmer and trimmed joists, although this is not so common in modern construction.

PRACTICAL TIP

If timbers that have already had a preservative treatment applied are cut to size on site then any fresh cuts need to have preservative treatment too.

PRACTICAL TIP

I-beams are used in designed floors and the drawing/schedule should be consulted in all cases. Do not make any alterations by cutting unless specified.

First floors (upper floors)

These are constructed in the same way as ground floors but deeper section joists are required to achieve the effective span without mid-span support. Upper floors will also require some form of strutting to prevent the joists from twisting.

Types of floor joist

Traditional floor joists are made from solid stress graded timber, while modern methods of construction use the I-beam, which is both stronger and lighter than its predecessor. The construction process remains the same with the exception of service installation. I-beams have pre-perforated circles, which can be knocked out with a hammer to allow the routing of cables and pipework.

CASE STUDY

LAING O'ROURKE

Working together

David Williams is the Regional Labour Apprentice Manager at Laing O'Rourke.

'All of the apprentices are assigned a mentor and a buddy. They form a bond by working together, and the mentor and buddy show the apprentice the ropes. Having the support of two people means there will always someone to talk to; someone to offer guidance, support and personal development.'

Installing floor joists

When installing different types of floor joists and strutting you need to follow any given specification for the job. It may also be necessary to form openings for service access, staircases and chimneys. The following practical tasks look at all of these activities.

PRACTICAL TASK

5. CONSTRUCT GROUND FLOORS

OBJECTIVE

To install ground floor joists and floors using different techniques, and to create flue and chimney openings.

Three methods are commonly adopted in domestic construction:

1.   Built-in with a clear span using bridging joists.

2.   Using sleeper or dwarf walls with wall plates, which reduces the section of the joists because of the reduced span.

3.   Using proprietary floor joist hangers.

Methods 1 and 3 can also be adopted for upper floors.

See previous practical tasks for details of the PPE requirements, and tools and equipment that you will need to complete this exercise.

> **PRACTICAL TIP**
>
> Consider the drawings, schedule or specification before starting the work. These will give the size, type and spacing of the joist to be used, any insulation requirements and the method of fixing will also be specified.

METHOD 1: BUILT-IN METHOD

Walls will be completed to DPC (damp-proof course) level.

> **STEP 1** Ensure that the inner blockwork walls are free of mortar and that the DPC is in place.

> **STEP 2** Measure across the supporting walls and cut the joists to length. Do not allow the joists to overhang into the cavity. (I-beams, if they are being used, will already be cut to size – see schedule and drawing.)

> **PRACTICAL TIP**
>
> Make sure that the joist ends do not extend into the cavity, which could lead to moisture penetration if the cavity is bridged.

> **STEP 3** Lay all the joists onto the walls camber (curve along its length) up.

> **STEP 4** Position the two outer joists 50 mm from the internal walls.

> **STEP 5** Set out the remaining joists at specified centres, usually 400 mm for solid joists and 600 mm for I-beams. Remember the penultimate joist centre can be under the specified centre but never over it

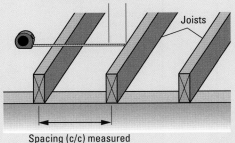

Practical spacing, measured from the inside of one joist to the outside of the next

Joists

Spacing (c/c) measured from centre to centre

Centre of joist

Figure 7.36 Set out joists to specified spacings

PRACTICAL TIP

Mark temporary battens with joist spacings in advance of Step 6, as this leaves both hands free for pulling joists to centre and nailing.

PRACTICAL TIP

The highest point may not be at one end – it could be in the middle, in which case work outwards in both directions to level the floor.

STEP 6 Use temporary battens (slate lathes/tile battens are ideal) to secure the joists on the centres and hold the joist in place for the bricklayer to wall the ends in. Leave the nails proud for ease of extraction later, and make sure the battens are clear of the walls at either end. The joists should be spaced and battened at both ends and in the centre depending on the span.

STEP 8 Level both the outer joists in their length and pack as before.

STEP 9 Attach a string line between the two outer joists on the opposite end to the end previously levelled and pack all remaining intermediate joists up to the line. Alternatively, use the straight edge and level as before.

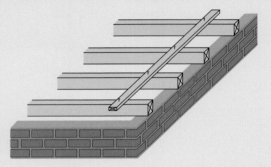

Figure 7.37 Use of temporary battens to secure joists before building in

PRACTICAL TIP

Joist ends should have some form of protection such as preservative, wrapping in DPC or proprietary plastic joist pockets.

STEP 7 Find the highest joist end – this is the starting point for levelling the floor. Now level from this joist, using a long spirit level and a straight edge. Continue along all the ends of the joists. Level to the furthest joist that can be reached with the straight edge, and pack with a non-compressible material such as slate. Then bring the intermediate joists up to the underside of the straight edge. Carry on until all joists are level on one side of the floor.

METHOD 2: GROUND FLOORS WITH SLEEPER WALLS/DWARF WALLS

This method is more costly but the joist ends are not exposed in the cavity. This eliminates the chance of moisture bridging into the joists and also allows complete air circulation.

The sleeper walls are built 50 mm from the internal blockwork. A DPC should be in place on top of the sleeper walls at both ends and on any intermediate honeycomb sleeper walls.

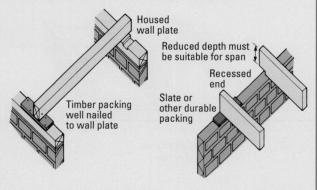

Housed wall plate

Reduced depth must be suitable for span

Recessed end

Timber packing well nailed to wall plate

Slate or other durable packing

Figure 7.38 Packing or recessing joist to maintain level

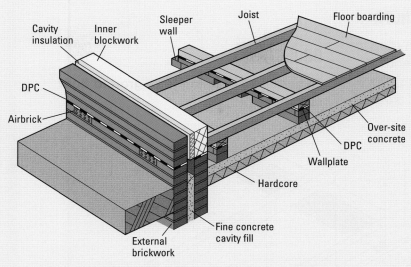

Figure 7.39 Timber suspended ground floor

STEP 1 Select wall plates from straight pieces of 100 × 50 mm timber. Cut these to length. It may be necessary to joint pieces in length; use a half lap header joint if longer lengths are required.

STEP 2 With the help of the bricklayer, bed the wall plates in mortar, tapping down until they are level both in their length and to each other. Check for parallel and leave until the mortar sets.

STEP 3 Measure the distance between the inner blockwork and deduct clearance.

Follow Steps 3 to 9 for the built-in method, above.

STEP 4 Using 75 mm round headed wire galvanised wire nails, skew nail the joists where they sit on the wall plates at both ends and on any intermediate wall plates. You may need to drill pilot holes.

PRACTICAL TIP

If care is taken to ensure that the wall plates have been levelled correctly then the packing stage of the process can be minimised.

METHOD 3: USING BUILT-IN JOIST HANGERS

STEP 1 Set out joist hangers to the required centres.

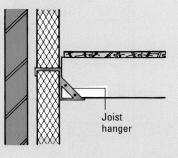

Figure 7.40 Built-in joist hanger

STEP 2 Measure the distance between the joist hangers. This should not be too tight as this may dislodge the upper row of blockwork, but it should also not be too short as this could allow the joist hangers to collapse inwards.

STEP 3 Cut and locate the floor joist. Repeat the levelling process as before, packing the joist hangers as required, and using temporary spacing battens as before. Do not pack the joist inside the hangers.

STEP 4 After the hangers are walled in, the joists can be nailed into the hangers to fill all the holes.

OPENINGS FOR GROUND FLOOR CHIMNEYS

In modern construction it is no longer common to trim out around fireplaces; however, carpenters working on maintenance or restoration projects will often have to repair or renew timbers around hearths.

OBJECTIVE

To install an opening for a ground floor chimney.

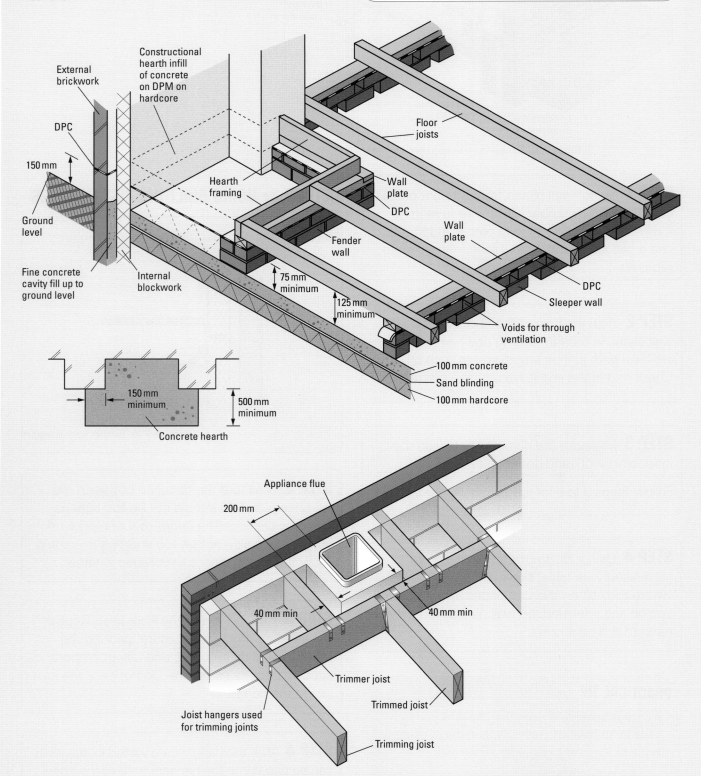

Figure 7.41 Openings for fireplaces and flues in suspended timber ground floors

STEP 1 Study the drawings, specifications and schedule; in the absence of the former the Building Regulations lay down guidelines for forming openings around chimneys and flues.

Around the hearth a fender wall is constructed in masonry; the floor joist will be trimmed around this.

STEP 2 Prepare and position all bridging or common joists as before.

STEP 3 Now cut and position the trimmed joist in front of the hearth; these sit on a wall plate bedded in mortar on top of the fender wall.

PRACTICAL TIP

The trimmed joists should not extend into the hearth.

PRACTICAL TASK

6. CONSTRUCT UPPER FLOORS

OBJECTIVE

To install upper floor joists and floors using different techniques, and to create flue and chimney openings.

Many of the operations used for upper floors are the same for ground floors; however, the joists will generally be deeper to allow greater spans in the absence of any mid-span support. Openings in upper floors are also considered in this section, such as for stairways and flues or services.

STEP 1 From the drawings, schedules or specification determine the method of fixing that is required (this will be either built-in or joist hangers).

STEP 2 Cut joists to length as before.

STEP 3 Where an opening is to be formed, start by measuring, cutting and positioning the trimming joists.

PRACTICAL TIP

Timber upper floor will consist of bridging joists, trimming joists, trimmer joists and trimmed joists.

STEP 4 Measure the distance between trimming joists to determine the length of the trimmer joist. Position the trimmer and fix according to schedule.

STEP 5 Set out the trimmed joists along the length of the trimming joist. The centres should be maintained once the bridging joists are positioned.

STEP 6 Measure, cut and fix trimmed joists to the trimmer.

STEP 7 Measure, cut and fix all bridging joists.

STEP 8 Fix temporary spacing battens as before.

STEP 9 Level all bridging, trimming, trimmer and trimmed joists.

STRUTTING

Look at the drawing and determine the method of strutting. There are two basic types: herringbone and solid strutting. There is a variation on timber herringbone strutting in the form of proprietary metal herringbone strutting.

HERRINGBONE STRUTTING

STEP 1 Mark out two parallel lines onto the top of any two joists in the positions that the strutting will be required. The spacing between the lines is equal to the depth of a joist minus 10 mm.

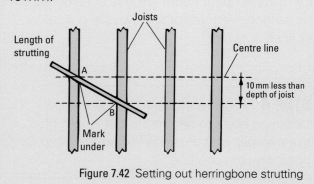

Figure 7.42 Setting out herringbone strutting

STEP 2 Hold the strut diagonally across the two lines. Mark the diagonal against the joist on the underside of the strut between the two adjacent joists. Offer the strut between the joists and check the fit.

STEP 3 If the strut fits and the gap between all joists is consistent then all other struts can be marked and cut in the same manner.

STEP 4 Make a saw cut at the top and bottom of each strut using a tenon saw. This cut is known as a kerf, which prevents the strut from splitting when nailed, but more importantly allows the struts to flex with any movement within the floor.

STEP 5 Using a chalk line, set the positions of each row of struts. Fix each row of struts in a cross pattern between adjacent joists using round-headed wire nails, making sure that the top and bottom of each strut does not stand proud of the joist surface. Finally, nail the two struts together where they cross.

STEP 6 Cut two sets of folding wedges for the ends of each row of struts. These should be hammered in between the wall and the last joist at each end of the run of struts. Secure by driving a nail through the joist into the wedges.

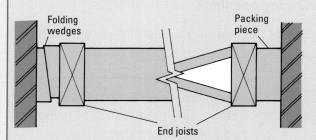

Figure 7.43 Folding wedges or packing piece to prevent bending of joist due to strutting

SOLID STRUTTING

Solid strutting can either be staggered or in line. Staggered strutting is easier to fix, but in line has the advantage of providing a continuous edge for fixing plasterboards if positioned correctly. The struts should be at least three-quarters of the depth of the joists. This type of strutting is inferior to herringbone strutting as it tends to become loose as the timbers dry out and services cannot be as easily run.

STEP 1 Use a chalk line to mark the position of each row of struts.

STEP 2 Measure, cut and fix the solid struts between each joist space.

STEP 3 Fix using 100 mm round-headed wire nails. Cut and fix folding wedges as before.

TEST YOURSELF

1. What is a MEWP?

   a. A working platform

   b. A type of steel used with floor joists

   c. A term used to describe certain floor openings

   d. A European law on health and safety

2. What can be used to hold up A-frames before the ridge board is nailed into position?

   a. Light blockwork

   b. A chisel

   c. Temporary braces

   d. Adhesive

3. What is a spreader bar?

   a. A tool for smoothing concrete

   b. A tool used to remove loose wood

   c. A PPE item used to protect the knees

   d. A tool used for the balanced lifting of trusses

4. Why is there a gap between the fascia board and brickwork in flush finished eaves?

   a. For ventilation

   b. To allow rainwater to escape

   c. Access for repairs

   d. Soundproofing

5. What is glue laminated timber also called?

   a. Plywood

   b. Glulam

   c. MDF

   d. Softwood

6. Which of the following materials are used to make binders?

   a. Concrete

   b. Steel

   c. Timber

   d. All of these

7. Which of the following is a bridging joist that has been cut off short to create a floor opening?

   a. Trimmed joist

   b. Trimmer joist

   c. Trimming joist

   d. Trimmable joist

8. What has largely replaced traditional trimming joints?

   a. Wall plates

   b. Tenon joints

   c. Joist hangers

   d. Struts

9. What should be the minimum thickness of solid struts?

   a. 18 mm

   b. 28 mm

   c. 32 mm

   d. 38 mm

10. What is the recommended gap between the outside joists and the wall?

    a. 20 mm

    b. 50 mm

    c. 75 mm

    d. 100 mm

Chapter 8

Unit CSA–L2Occ39
CARRY OUT MAINTENANCE TO NON-STRUCTURAL CARPENTRY WORK

LEARNING OUTCOMES

LO1/2: Know how to and be able to prepare for maintenance operations

LO3/4/5: Know how to and be able to repair timber and timber-based components

LO6/7: Know how to and be able to replace guttering components

LO8: Know how to replace sash window cords

INTRODUCTION

The aims of this chapter are to:

* show you how to repair timber components

* show you how to replace guttering systems

* show you how to replace sash cords

* show you how to reinstate surfaces.

PREPARING FOR MAINTENANCE OPERATIONS

It is rare for domestic dwellings to have clear or regular maintenance programmes. Commercial buildings, or those that are owned by housing associations and local authorities, often have set maintenance schedules. When a construction company is asked to carry out repairs and maintenance they will usually survey the whole building. This will give them the chance to identify the nature and extent of any work that needs to be carried out.

This chapter does not look at major structural repair and maintenance work. It focuses on non-structural carpentry and related tasks. In many cases the maintenance work will come about as a result of the occupier of the building noticing a defect in a certain area.

Potential hazards

Many of the potential hazards that were relevant in first and second fix and other more specialist work are relevant. You should refer back to Chapter 1 for information on how to prevent problems with health and safety and carrying out risk assessments.

Schedules and specifications

The initial survey to identify defects usually begins by looking at the outside of the building. Damp patches in brickwork could suggest that there might be a problem with the guttering or downpipes. Other external problems may be more obvious, such as a cracked pane of glass or rotten window cill.

Once inside the building, there may be a variety of different problems. Visible holes in woodwork could suggest an insect infestation. A door may not close properly. This could be a problem with the hinges, ironmongery or the door may be twisted or actually swollen. This might also be the result of movement within the structure or subsidence.

The normal course of action is to:

* identify the location of the defect

* note its possible cause

* suggest remedial action to correct the defect.

This process will identify a clear way forward. It will enable a list to be compiled of the work to be carried out, and prioritise the defects, ensuring the most serious is completed first. This will form the basis of a schedule of work, suggesting how long the work will take (and probable cost). Finally a specification can be completed listing the detail and quality expected. It will also help to state the level of work or specification that will need to be done. Method statements and risk assessments will also be generated from the survey.

PPE

In Chapter 1 we looked at PPE in relation to collective protective measures and personal protective equipment. We also looked at respiratory protective equipment, such as a LEV, which are useful in enclosed spaces.

Timber used for mouldings, doors and window frames

Mouldings, doors and window frames are normally made from softwood or artificial composite materials:

* Softwood – good stockists will have a large range of softwood mouldings and profiles ready for use. Timber can be bought in any desired stock size from 1.8 m, going up in 300 mm stages to 5.4 m. Mouldings are normally purchased in linear metres for smaller amounts. Square stock such as 75 × 50, 100 × 50, etc. would normally be bought in cubic metres for large quantities. Softwoods can come from managed forests and are usually redwood or whitewood. Douglas fir softwood would normally need to be specified as it is not often available off the shelf. These species vary considerably in both quality and durability.

* MDF – this is being increasingly used for mouldings, windows and boards. It comes in a variety of profiles, sizes and finishes. It can come ready-primed, veneered or laminated.

* Chipboard – this is very versatile and lightweight. It is used to construct doors, not mouldings.

* Plywood – this comes in various board sizes, usually 2.44 m by 1.22 m, and can be up to 38 mm thick. The veneers are at right angles to give stability to the board. Plywood can also be curved.

Figure 8.1 A local exhaust ventilation system

Figure 8.2 Dry rot

Figure 8.3 Wet rot

KEY TERMS

Biocide

– this is a chemical substance that can kill living organisms and spores.

Wet and dry rot

When timbers have a moisture content of more than 24 per cent they are more susceptible to wet rot. It is caused by direct contact with water, such as leaks from gutters and pipes.

Wet rot is easy to notice because it encourages the growth of a fleshy mould. The strands of the mould are usually classed as either brown or white. This classification does not refer to the colour of the mould, but to what it does to the timber. Brown rot makes wood darken and crack. White rot affects the grain and causes the timber to lighten in colour. Both make timber weak and in most situations you will be able to pull pieces of timber away using your hand.

Wet rot can be managed in the following way:

* Find the source of the dampness and cure the defect.

* Dry out the surroundings and remove damaged timbers, as they will no longer be strong enough.

Dry rot is more likely to occur in timbers with a slightly lower moisture content of between 18 and 22 per cent. It is caused by high moisture content coupled with a lack of ventilation, which results in high humidity.

Dry rot can actually travel through masonry in order to find more timber, for example the structural timber. Dry rot spores have the appearance of cotton wool (known in the trade as 'dead man's fingers'). They break down the lignin in the wood and they will extend along timber or masonry. Dry rot will eat any nutrients it finds in the timber. It can be managed by:

* determining the source of the moisture and curing the defect

* cutting out and removing any affected timbers and removing up to 600 mm of the timber beyond the visible signs

* stripping the plaster back until there are no visible signs

* drying out the area if possible

* cleaning any surrounding masonry with a steriliser

* drilling and injecting a **biocide** into thick walls

* spraying any surrounding timbers with the biocide.

Insect infestation

It is important to understand the development of insects that can affect timber. In fact some of the young (larvae) can live in wood for ten years.

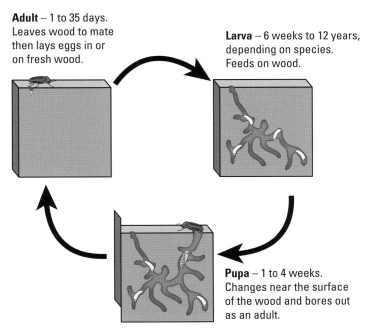

Adult – 1 to 35 days. Leaves wood to mate then lays eggs in or on fresh wood.

Larva – 6 weeks to 12 years, depending on species. Feeds on wood.

Pupa – 1 to 4 weeks. Changes near the surface of the wood and bores out as an adult.

Figure 8.4 The life cycle of a wood-boring insect

The majority of damage done to timber in the UK can be attributed to five species of insect:

* common furniture beetle

* powder post beetle

* death watch beetle

* house longhorn beetle

* weevil.

Often referred to as woodworm, it is the larvae of the insect that are responsible for the damage.

Figure 8.5 Woodworm holes

As with rot, the ideal solution is to remove the conditions that allow the insects to thrive. The main thing to do is reduce the moisture content of the timber. Any rotten timbers are more likely to be attacked by insects, so this means looking for leaky roofs and pipes and poorly maintained guttering.

Woodworm

Woodworm is a generic term for wood-boring insects. They create small holes and leave egg-shaped pellets in the dust they create. The larvae can live for three years and will bore into the wood. They can cause huge damage to structural timbers. If the moisture content of the timber is less than 12 per cent it is hard for the larvae to survive but if the moisture content of the timber is high and there is existing rot then 'woodworm' can add to the structural failure of the timber.

Softwood, such as pine and fir, is particularly vulnerable to insect infestation. Usually it is necessary to replace structural timbers that have been affected.

Figure 8.6 Flight holes caused by the common furniture beetle

Figure 8.7 Larvae of the powder post beetle

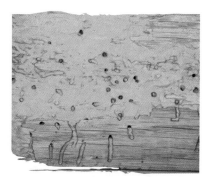

Figure 8.8 Flight holes caused by the powder post beetle

Figure 8.9 A death watch beetle

Common furniture beetle

These are the most common form of wood-boring insect. They measure 2.7 mm to 4.5 mm in length and have a brown body. The larvae will attack soft and hardwoods. Their natural habitat is broken tree branches and areas where tree bark has been damaged or removed. They will attack any type of woodwork including flooring, joinery and structural timbers. They do not like very dry wood, so are less likely to be seen in timbers that are kept dry by heating systems, preferring areas of the house where there may be some dampness. The flight holes are around 2 mm. The insects will bore deep into the wood and create many tunnels and galleries.

Powder post beetle

These will attack most types of new timber, including flooring. They produce a fine dust and holes 2 mm in diameter.

The flight holes can be seen clearly in Fig 8.7. The dust around the holes is called frass and is the waste material left by the activity of the larvae and the beetles' flight from the timber.

Death watch beetle

This wood-boring insect is related to the common furniture beetle, but is much larger. Death watch beetles are approximately 7 mm long with larvae growing up to 11 mm long. They prefer very damp conditions, especially when there is some kind of fungal decay such as wet rot in the timbers. They tend to attack timber that is already rotting or decaying but prefer European hardwoods, especially oak, ash and chestnut. Their larvae can live for ten years. They are often found in hard-to-see places, where the ventilation is poor. If they are allowed to establish themselves, the timber will fail. They leave holes around 3 mm in diameter.

The house longhorn beetle

These are about 25 mm long. House longhorn beetles are typically black or brown. They are principally found in roof timbers, where they attack the sapwood of exclusively softwood species, often resulting in severe structural weakness. The holes and tunnels of the house longhorn beetle are much larger than those of the furniture beetle. The damage caused by these insects can be severe.

Treatment

If an infestation is found, then it is wise to seek specialist advice. They will be able to advise you whether the infestation can be treated without having to replace affected timbers or whether the timbers will need to be taken out. The opinion of the specialist may have an impact on the work you are doing and whether it needs to be delayed as a result of necessary treatments.

It is sometimes necessary to use chemicals to treat the infestation. Usually, this should be carried out by a specialist. There are several options:

* Spraying – this is good against furniture beetles, but as the spray is only applied to the surface of the wood it may not be effective against those that are deep inside the timber.

* Injection – small holes are drilled into the timber, or into the flight holes of the insects and a chemical injected. The chemical penetrates deep into the timber and kills the majority of the insects.

* Paste – this is a good alternative to spraying, as the chemical will slowly penetrate into the timber.

* Smoke – this will usually kill adult insects that are emerging from the timber, but will have virtually no effect on eggs or larvae. This means that the treatment has to be repeated over several years to kill off all the insects.

* Fogging – this is a chemical fog that is used when large areas are affected. The chemicals target the eggs. The problem is that the deeper the eggs are in the timber the less likely they will be affected by the chemical fog.

Protecting work and its surrounding area

Even though non-structural carpentry maintenance may not be as extensive or disruptive as structural repairs it is still important to ensure that you minimise damage, maintain a clear workspace and dispose of any waste produced.

More information on protecting the work and its surrounding area can be found in Chapter 4.

REPAIRING TIMBER

Timber products can fail over time. There are various reasons for this:

* Faulty design or construction – the building may have been constructed using poor specifications or poor quality materials and components. This may cause the building to move, suffer moisture penetration or biological attack.

Figure 8.10 Damage caused by damp

Figure 8.11 Chemical damage caused by metal

* Dampness – this can be caused by rain penetrating the building, condensation, leaky plumbing or heating, rising damp or a failure, such as a bath overflowing or leaking.

* Movement – cracks and gaps may appear as a result of the ground underneath the building settling. Also, building materials can move as a result of changes in moisture, temperature or chemical changes.

* Chemical and biological attack – metal can, of course, corrode and this can affect brickwork, iron and steel, and make timber vulnerable. Wood-boring insects and fungi can exploit the weaknesses in the timber.

* Weathering – over time certain timber that is exposed to the elements will weather and become damaged as a result of wind, frost, snow, sun and rain.

* Fire – timber is combustible, which means it is capable of burning. But even a fire close by can scorch and damage woodwork. Timber can be discoloured by smoke. Although structural timber will burn at a slow rate, its structural strength can be affected and this could lead to a collapse.

Disposing of affected timber

It may be impossible even to try to recycle timber that has been badly affected. This is certainly the case when dealing with timber that has had wet or dry rot. It needs to be completely removed from the building.

Normally it is taken away and burned. Do NOT start open fires on site – if this timber is to be destroyed it must be taken away. The danger of leaving any affected timber, for example, under the floorboards, is that the rot will spread once again into the replaced timbers.

Figure 8.12 Weathered wood

Paint systems for timber components

Woodwork is often painted or varnished to provide a good decorative finish. This will protect it from the elements. In most cases a timber is knotted and then a primer is used to seal the surface. This has two purposes:

* It provides a bond for further coats of paint.

* It seals the surface of the wood.

The primer is then covered with a second or even third surface of paint, which gives a good overall coat either in gloss or satin.

Increasingly, water-based paints are being used rather than oil-based paints. This is as a result of the drive towards greener, more environmentally friendly products. The chief environmental issue

Figure 8.13 Smoke damage to wood

confronting the paint industry is that of volatile organic compounds (VOCs) emissions and their impact on air pollution, that is, the effect of solvents in paints on the air that we breathe.

There were concerns in the past about the durability of water-based paints, particularly for external use.

There are various steps that can be taken when painting surfaces, as can be seen in Table 8.1.

DID YOU KNOW?

Paints are either water-based or solvent/oil-based. Solvents include white spirit and turpentine. They contain VOCs (volatile organic compounds), which can lead to allergic reactions and illness. The oils and solvents are also made from non-renewable resources and are pollutants. Environmentally friendly paints are mainly VOC free and use natural ingredients.

Bare wood	Previously painted wood	Small areas
1. Clean down with white spirit	1. Clean off using sugar soap	1. Remove loose paint
2. Knot	2. Rub down and remove any loose paint	2. Rub down
3. Prime	3. Apply knotting solution to any knots that are bleeding through	3. Treat any small areas with wood hardener
4. Stop (fill)	4. Prime any edges or bare timber	4. Undercoat
5. Rub down	5. Rub down	5. Stop (fill)
6. Undercoat	6. Undercoat	6. Undercoat
7. Stop (fill)	7. Stop (fill)	7. Topcoat
8. Rub down	8. Undercoat	
9. Undercoat	9. Topcoat	
10. Topcoat		

Table 8.1

It is important to make sure that you follow careful procedures, whether you are carrying out internal or external painting. Table 8.2 suggests some precautions that you should take in both cases.

Internal painting precautions	External painting precautions	
Cover any furniture or carpets with a dust sheet.	Never work in strong sunlight. The paint may dry patchy and may blister.	
If you are rubbing down then seal the doors to other rooms with masking tape and open the window.	Make sure that the woodwork is dry. Do not paint if you expect rain.	
Wear a dust mask and goggles if you are scraping paint off or rubbing down.	Do not paint first thing in the morning or late in the afternoon as there may be moisture in the air or on the woodwork.	
Always apply paint along the grain of the wood. This applies to primers and undercoats.	Never paint if there might be a frost.	
To give an even paint coverage, apply top coats along the grain then brush it out across the grain and finally give a brush stroke across the grain.	Make sure you do not use the wrong paint, so check that the paint has not been designed for internal use.	

Table 8.2

Figure 8.14 shows the order in which you should paint doors and windows. This helps to blend areas of paint together (keep a wet edge) so that when the paint dries there are no obvious joins.

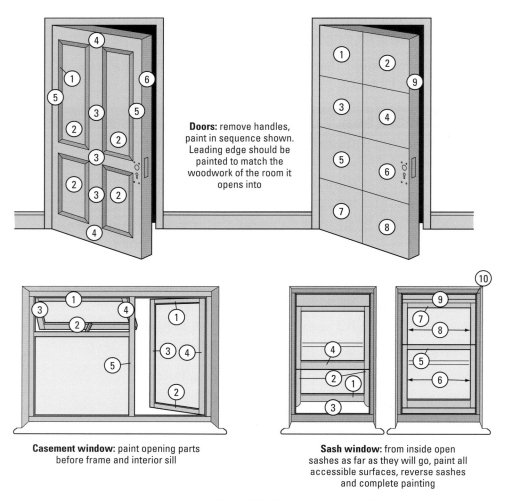

Doors: remove handles, paint in sequence shown. Leading edge should be painted to match the woodwork of the room it opens into

Casement window: paint opening parts before frame and interior sill

Sash window: from inside open sashes as far as they will go, paint all accessible surfaces, reverse sashes and complete painting

Figure 8.14 Sequence for painting doors and windows

Applying preservatives

Any wood preservatives that are used on timber were formerly approved under the Control of Pesticides Regulations (1986/1997) but they were replaced in September 2013 by the EU Biocides Regulation. Only people that have been trained and are competent to use them are allowed to apply preservatives.

The amount of protection that a wood preservative gives to the timber is dependent on:

● the depth to which the preservative has penetrated the timber

● how long the preservative is expected to be effective.

Timbers that are exposed to the elements, particularly timbers that will be exposed to rainwater and changes in temperature, present a higher decay hazard than those used for indoor woodwork. Any exterior timber that has contact with the ground is likely to decay far quicker.

PRACTICAL TIP

The Control of Substances Hazardous to Health (COSHH) Regulations also apply to the use of wood preservatives.

The simplest method of applying preservative is to paint it on. Soft bristled brushes are used to provide a flood coat. The timber needs to be clean and dry. Subsequent coats can be added once the previous coat has soaked in, but has not dried out. The preservative needs to be flooded into joints and on end grain for better protection.

Another method is to apply the preservative as a paste. This technique is usually employed where access is difficult.

A third option is to use high or low pressure treatment. The timber is loaded into a cylinder and then the cylinder is filled with the preservative. A pressure is then applied. The gauges on the cylinder will show just how much of the preservative has been absorbed by the timber. This is a technique that is often used to apply **creosote** to timber.

Two of the most common methods of pressure treatment are Tanalizing and double vacuum treatments, using fire retardant preservatives. A relatively new timber product called Accoya is treated with acetic acid and has an estimated life span of 50 years.

A final alternative method of preservation is to inject the timber. Holes are drilled into the timber and a strong fungicide and insecticide gel is injected into the holes. This is available in a cartridge and can be applied using a sealant gun. The holes are then plugged with dowels.

Water-based preservatives

Water-based preservatives are usually quite effective for exterior woodwork. They are anti-fungal, but they are not as effective as waterproof preservatives. In many cases water-based preservatives are only used on interior wood, but not in areas where the woodwork may be exposed to water, such as kitchens or bathrooms.

Spirit-based preservatives

Spirit-based wood preservatives are good for overall external use. They are waterproof. Usually the preservative needs to be reapplied from time to time to keep it effective.

Access equipment

When you are working on repairing timber it is likely that you may need access equipment and be working at height. This means that the Work at Height Regulations (2005) apply.

In Chapter 1, there is information about the Work at Height Regulations and how to ensure that any equipment and safety measures you use will protect you. You should always try to avoid working at height if possible. If it cannot be avoided, always use the proper equipment and put measures in place to prevent falls. There will always be a risk, even if the proper equipment and safety measures are taken. As there is always a risk, measures need to be put in place to minimise the distance and consequences of a fall.

Figure 8.15 Applying preservatives

Repairing timber mouldings

This practical task shows you how to select the relevant hand tools, as well as how to replace damaged timber mouldings. It also shows you how to splice new sections of mouldings as required.

PRACTICAL TASK

1. REPAIRING TIMBER DOORS AND WINDOW FRAMES

OBJECTIVE

To use different methods to repair damage to doors and windows.

SPLICING

Splicing is the term used to describe the letting in of a new piece of timber to repair a damaged component, such as a stile on a sash or a jamb on a door frame. Three variations are described below.

TOOLS AND EQUIPMENT

Hand tools:

Tape measure	Claw hammer
Pencil	Screwdrivers
Spirit level	Drill bits
Straight edge	Countersink
Handsaw, panel and or tenon	
Chisels	Combination square
Marking gauge	

Power tools:

Hammer drill	Jig saw
Cordless screwdriver	110V transformer
Chop saw	Router
Drill bits, for wood and masonry	

PPE

Ensure you select PPE appropriate to the job and site where you are working. Refer to the PPE section in Chapter 1.

METHOD 1: This could be used on the top or bottom of a door stile, window jamb or a window cill and many other applications.

STEP 1 Mark out the area to be replaced using a 45° combination square. The repair is angled to act as a key for the new timber. Cut the damaged area in an upward direction. Mark the depth of the repair along its length using a marking gauge.

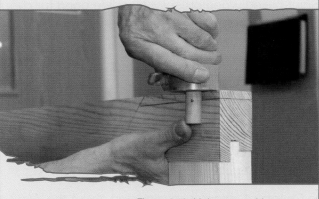

Figure 8.16 Using a marking gauge

STEP 2 Remove the damaged timber using a tenon or panel saw or a chisel, as required, working back to the gauge lines.

STEP 3 Cut the new piece of timber slightly longer, wider and thicker than the recess in the component.

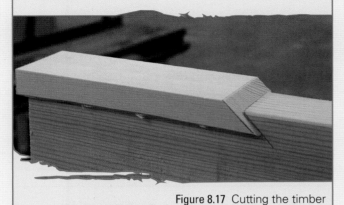

Figure 8.17 Cutting the timber

STEP 4 Treat the area being repaired with a suitable timber preservative.

STEP 5 Drill and countersink the repair piece and fix using glue and screws.

Figure 8.18 Fixing the repair piece

STEP 6 Dress the spliced timber to the component size using planes.

Figure 8.19 Dressing the timber to size

STEP 7 Use a filler knife to fill the screw holes and any minor gaps in the repair. Leave the filler to set.

STEP 8 Sand the repair to a smooth finish.

STEP 9 Apply a primer coat, two undercoats and gloss to finish.

METHOD 2: This method can be adopted when a repair is towards the centre of a component.

STEP 1 Mark the area to be removed with 2 × 45° angles at either end of the repair. Mark the depth along the length of the repair with a marking gauge.

Figure 8.20 Marking the area to be removed

STEP 2 Remove the damaged area by making the 2 × 45° cuts down to the gauge lines. Then make a series of intermediate cuts, again down to the gauge lines.

Figure 8.21 Cutting to 45°

STEP 3 Chop out the recess using a sharp bevel edged chisel; this should be the widest chisel that the work will allow. Ensure you work from the outside edges to the centre, to avoid splitting out.

Figure 8.22 Chopping out the recess

STEP 4 Mark the new piece of timber that is to be spliced in. This should be the exact length at the narrowest points between the mitres, and slightly wider and deeper than the recess. Cut using a tenon saw.

STEP 5 Treat the area being repaired with a suitable timber preservative. Drill and countersink the repair piece and fix using glue and screws, then dress the spliced timber to the component size using planes. Sand the repair to a smooth finish.

Figure 8.23 Dressing the timber

STEP 6 Apply a primer coat, two undercoats and gloss to finish.

METHOD 3: This is similar to Methods 1 and 2 but the difference is that the repair does not extend to both faces of the component.

STEP 1 Mark out as in Method 1 or 2; use a marking gauge to mark both the depth and the width of the splice in this instance.

STEP 2 Make a series of angular cuts to the gauge lines – this includes the 45° cuts.

STEP 3 Using a sharp bevel edged chisel remove the waste in two directions. Feather the recess before removing the waste across the grain.

Figure 8.24 Removing the waste

STEP 4 Mark out and cut a new piece as for previous methods.

STEP 5 Follow Steps 5 and 6 as above.

PRACTICAL TIP

Protect the surrounding area with dust sheets if necessary. Rotten timber should be removed from site immediately and ideally burned. Undamaged joinery in the immediate vicinity of the repair should be protected with MDF, plywood, corrugated cardboard or similar to prevent damage.

CASE STUDY

South Tyneside *Homes*

South Tyneside Council's
Housing Company

It feels great to do a good job

Glen Campbell, now a team leader, completed his apprenticeship at South Tyneside Homes.

'Working in the maintenance team was something I really enjoyed because it gave you the chance to develop your skills slowly and steadily on a one-to-one basis with a tradesperson. The jobs weren't massively taxing when I look at them now, but at the time, to be able to do them little bit by little bit, without having a strict time frame put on you, was much better than working in a pressured environment where it's got to be done in half an hour because you've got to make money.

Joinery is an aesthetically pleasing trade. As a joiner you're there to hide things a lot of the time or to dress it. Everything that you do joinery wise, apart from your roofs, is there to be seen, so what you are doing has to have a good finish. It's got to be pleasing to the eye, especially your walls because you wouldn't want to sit in a house with a wall that's all wavy, reflecting the light and refracting all over the place with scratches and marks. It's about getting the finish your customers want, but if you're a good tradesperson, it's also getting the finish that you want.'

REPLACING GUTTERING

As we will see, some guttering can be made from cast iron or aluminium. A carpenter or plumber will often be the person who needs to replace gutters and downpipes, along with other maintenance activities. Sometimes if a carpenter is working at this height the task may fall to them.

Gutters are designed to collect rainwater coming off the roof. They are positioned in such a way as to fall in the direction of the downpipes. It is essential that they are kept free of debris to allow effective operation.

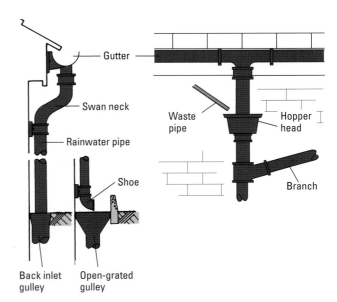

Figure 8.25 Rainwater pipes and connections

Materials used for gutters and downpipes

The vast majority of modern buildings have plastic gutters and downpipes.

Aluminium gutters are available in both cast and extruded forms. The extruded form is formed on site from a roll of flat aluminium that is mounted in the back of a vehicle. Various profiles can be chosen, and then the flat aluminium is passed through the chosen profile to produce a continuous length of gutter. This is called cold forming. The advantages of this type of guttering are that it is seamless, resulting in fewer leaks, and requires no brackets – it is simply fastened through the back directly to the fascia board or rafter ends. There is also no waste as the sections are made to the exact measurements. However, its main disadvantage is that it dents easily if a ladder is lent against it.

Cast aluminium guttering and fall pipes are made to look like cast iron with the advantage of being much lighter. It is installed in much the same way as cast iron guttering, using brackets and connector bolts.

Older properties may have had their original cast iron guttering and rainwater pipe systems replaced with plastic ones. However there are still many, particularly buildings that are protected (e.g. listed buildings or those in a town's conservation area), which are required to retain their old systems. This means it is possible to come across not only cast iron systems but also lead, copper and even asbestos cement. Timber guttering was still routinely used until the 1950s, even on new build. Plastic guttering systems began to be used extensively after this because the development of plastics sped up during the Second World War.

Stone gutters were used on the oldest buildings, such as churches, cathedrals and castles, and later lead was used. The dissolution of the monasteries in the sixteenth century resulted in an abundance of lead being available because it was taken from those buildings and this was the material of choice for a while.

Concrete gutters were used extensively after the Second World War on what became known as pre-fabs – sectionalised concrete housing – many of which are still standing today.

Guttering system components

The parts of a guttering system are designed so that they can be fixed together to fit any kind of shape or length required. Some of the fittings are designed to join longer sections together. Other parts are there to end or terminate the gutter, or to support it.

PRACTICAL TIP

If you come across gutters or downpipes that you believe are made from asbestos cement you should stop work immediately, report it to your supervisor and get the necessary advice on how to deal with it.

Gutter component	Description and purpose
Pipe	Rainwater pipe (RWP), downpipe or fallpipe. This refers to the vertical tube that makes up the downpipe.
Shoe	This is a fitting that is used on a downpipe to change the direction of the downpipe by around 45°. It directs the water into the rainwater gulley.
Joint	Each time a component of the gutter or downpipe is connected it is necessary to have a joint or fixing. Many of the components are designed to slot into joints on other components or, if necessary, jointing brackets or unions can be used.
Hopper	This is also known as a hopper head, leader head or conductor head. The rainwater collected by the gutter feeds into the hopper and then passes down the downpipe and into the drain.
Elbow	This is an angled part of the piping. It is attached to the down spout. An elbow can be square or rounded and it is used to direct water flow into a hopper head.
Offset	This is a pipe that directs a downpipe from the gutter. It runs under the soffit of the eaves and down the wall.
Gutter	This is a semi-circular or other shaped channel that collects the rainwater that falls onto the roof.
Clips	These are used at most joints and connectors to hold the guttering in place. They are usually wing-shaped.
Brackets	These hold the guttering or the downpipe to the building.
Running outlet	This provides an outlet to the downpipe on the length of the guttering. This means that there is guttering at either side of the running outlet.
Stop end	This closes off a run of guttering.
Stop end outlet	This closes off the end of the gutter and also provides an outlet to the downpipes.
Angled bends	These allow the run of guttering to continue around corners.

Table 8.3

Guttering and downpipe profiles

There are several different styles and shapes of guttering and downpipes. The earliest ones were box-like channels. When metal guttering was introduced the half-round shape became the most common. Today, with new manufacturing techniques, other shapes are possible.

Square
Square gutters and matching downpipes are usually made from plastic. They are available in a variety of colours, including white, brown, grey and black.

Round
Different shaped guttering can be fitted with round downpipes. These are also available in a range of colours and materials, from traditional cast iron through to lightweight and durable plastics.

Half round
Half-round guttering is perhaps the most common type that is used today. It is usually matched with round downpipes.

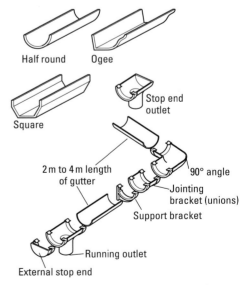

Figure 8.26 Guttering profiles and fittings

Ogee

This is a curve that consists of a pair of arcs that bend rather like an 'S' shape. Both of the ends are parallel. It is an old-fashioned style, which was introduced around 150 years ago.

Guttering joints

The way in which guttering is jointed will depend on the material being used. Metal guttering sections, whether they are made from cast iron or steel, are sealed at their joints and then bolts are put through for further strength. Over time the bolts can become corroded. Each of the joints will have a layer of roof and gutter sealant and the actual gutter sections overlap. The bolt holds the two sections together.

More modern plastic or PVC guttering fixes together using sections called union pieces. These union pieces have a rubber gasket. The union pieces have lips that hold it in place. These can be simply replaced or new gaskets can be fitted into the joint.

Figure 8.27 A leaking gutter

Identifying damage to gutters and downpipes

Leaking gutters are one of the major reasons why there might be isolated damp patches in a building. Sometimes it is not immediately obvious that water is leaking from the gutter. If left the water will stain the wall and there is a danger that moss and algae will grow.

Gutters can also overflow. This usually means that there is a blockage. The usual reason is that there is natural debris that has blocked the gutter. This could be moss from the roof or leaves. Other common reasons for overflowing guttering are objects that have either fallen onto the roof and into the guttering, or objects like tennis balls that have got stuck there. The water will spill out over the side of the gutter when there is a blockage.

Figure 8.28 A gutter with a leaf guard

One way of avoiding many of the problems of debris in gutters is to fit a leaf guard, or plastic grill. These can be purchased to fit most modern guttering, although in older properties chicken wire can be cut and bent to cover the top of the guttering. This would need to be secured with cable ties or wires.

Gutters and downpipes can also suffer damage in extreme weather conditions. If there is a blockage then any trapped water will expand as it freezes. This can cause downpipes to crack or joints to fail. In older properties that have cast iron downpipes corrosion is also a problem.

In all cases, metal fittings for both gutters and downpipes will fail over a period of time. They will corrode and can become loose. This will allow movement that could put pressure on joints. Lead and cast iron downpipes can actually crack if the protective paint layer breaks down.

PRACTICAL TASK

2. REPLACE GUTTERING COMPONENTS

This practical shows you how to replace damaged guttering systems and joints in guttering and downpipes. It may be necessary to erect a working platform when replacing gutters and rainwater pipes, as ladders are not always the best way of accessing the area. In most cases there is a standard procedure that should be followed. This can be seen in the following diagram.

TOOLS AND EQUIPMENT

Tape measure	Chisels
Hacksaw	Claw hammer
Screwdriver, hand or cordless	
Drill bits	String line
Nail-bar	Hammer drill
Masonry bits	110V transformer

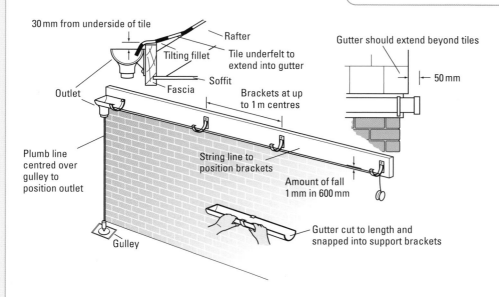

Figure 8.29 Replacing guttering

OBJECTIVE

To replace and repair plastic and timber guttering components in accordance with individual specifications.

Although rainwater systems can be formed from plastic, cast iron, timber, aluminium, concrete and asbestos, the following methods cover the most common types – plastic and timber rainwater systems (often referred to as rainwater goods on drawings and specifications).

PPE

In this, and the tasks that follow, ensure you select PPE appropriate to the job and site where you are working. Refer to the PPE section in Chapter 1.

PRACTICAL TIP

Gutters can be fixed level or have a fall of 1:600 or 10 mm in every 6 m.

3. REPLACE PLASTIC (PVC) RAINWATER GUTTERING AND DOWN PIPES

STEP 1 Carefully remove all guttering, brackets and down pipes, taking care not to damage the bottom row of slates or tiles.

Figure 8.30 Remove old gutter

STEP 2 If necessary repair or repaint the fascia board at this stage.

STEP 3 Fix the gutter outlet to the fascia board 50 mm below the tiles and in line with the rainwater gulley below. The roofing felt should reach the centre of the gutter outlet.

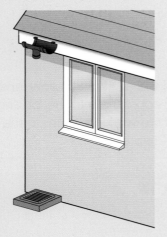

Figure 8.31 Fix outlet in line with rainwater gulley below

STEP 4 Fix a gutter bracket at the other end of the gutter run; this should be as tight as possible under the tiles or slates while still allowing the gutter to be clipped in. This will provide the necessary fall to the gutter outlet.

Figure 8.32 Fix a gutter or fascia bracket approximately 150 mm from end of gutter run

STEP 5 Attach a string line between the gutter outlet and the gutter bracket. Ensure this is taut and touching both at the same point.

STEP 6 Fix all intermediate gutter brackets at the manufacturer's specified centres.

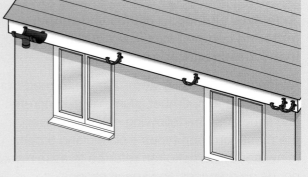

Figure 8.33 Fix intermediate gutter brackets

STEP 7 Fix the stop end to a length of gutter. Make an allowance for expansion and contraction; usually there is a mark on the inside of the fitting that the gutter should not exceed.

Figure 8.34 Fix a stop end to gutter. This can be done before or after the gutter is in place

STEP 8 Cut the gutter to length or alternatively use a jointing bracket if the gutter run is longer than a single length. Take care to ensure the gutter sits within the expansion lines at either end.

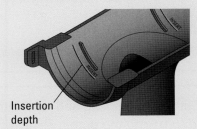

Insertion depth

Figure 8.35 The gutter must always line up with the insertion line on any fittings

PRACTICAL TIP

Offsets are often made in one piece to certain angles and lengths; however they can be made up from fittings and a length of downpipe.

PRACTICAL TIP

Downpipes or fall pipes that transmit water from the gutter to the rainwater gully should not be confused with drainpipes, which are found underground.

STEP 9 Start at the outlet end and clip the gutter into the brackets. When clipping into outlet gulleys and jointing brackets, make sure the rubber seals are properly seated within the fittings.

STEP 10 Cut the new downpipe to length. It should be fitted plumb over the rainwater gulley or drain; a suitable offset should be used to span the eaves overhang.

Figure 8.36 Offsets are required to bridge the fascia and the house wall

STEP 11 Fix the downpipe clips directly behind the centre line of the pipe using zinc round-headed sheradised or brass screws and plugs at 1.8 m centres.

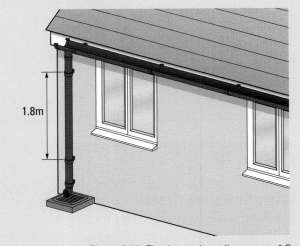

1.8m

Figure 8.37 Fix downpipe clips every 1.8 m

STEP 12 Fix a downpipe shoe to the bottom of the pipe. This is often incorporated in a bracket.

PRACTICAL TASK

4. REPLACE TIMBER GUTTERING

Timber guttering can be fixed in a number of ways:

* nailed directly to the fascia board
* sat on brick corbels and fixed directly into the brickwork seam with iron dogs
* secured to the rafters with iron gutter brackets in roofs with open eaves.

STEP 1 Take down all rotten guttering, being aware that timber guttering is very heavy especially when it is rotten, as it becomes waterlogged. It may be necessary to gently lever the guttering away using a nail-bar. De-nail all the old guttering.

STEP 2 Cut a new piece of gutter to length. If the gutter run is longer, a lapped butt or heading joint will have to be formed to increase the overall length. The lapped heading joint or butt joint is formed by cutting along the quirk on the underside of the gutter.

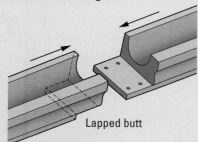

Lapped butt

Figure 8.38 Lapped heading joint in timber gutters

PRACTICAL TIP

In some situations slates or tiles at the eaves may have to be removed in order to remove gutter brackets. It is often more cost effective to grind off the old gutter brackets and replace them with brackets that fix to the side of the rafters. These may have to be fabricated.

STEP 3 Treat the joint with preservative.

STEP 4 Using a sharp chisel, recess the gutter into both halves (to take the piece of lead in Step 5). Bring the joint together and seal with bitumen mastic, then fix from underneath using screws.

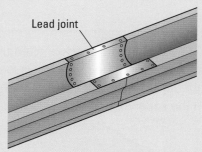

Lead joint

Figure 8.39 Lead recessed into gutter across lapped joint

STEP 5 Cut a piece of lead and lay it into the recess and form into the gutter. Fix it with copper nails.

PRACTICAL TIP

Use a lapped mitre joint on hipped or valley roofs or gutters with a stopped return around a gable end. The joint is formed as in Step 5.

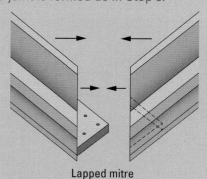

Lapped mitre

Figure 8.40 Lapped mitre joint

5. PREPARING THE GUTTER FOR THE DOWNPIPE OUTLET

STEP 1 Using an expansion bit, drill a hole for the lead funnel to pass through.

STEP 2 Recess the bottom of the gutter to the size of the flange on the funnel and treat and fix as before.

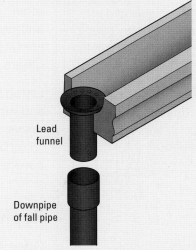

Lead funnel

Downpipe of fall pipe

Figure 8.41 Lead funnel recessed into gutter

PRACTICAL TASK

6. REPAIR TIMBER GUTTERING

Although timber guttering is rarely used in modern construction, many heritage projects and repairs to older properties require the carpenter or joiner to have the knowledge and skills to execute this type of work. The methods used for replacing timber guttering should be adopted for repairs but will have to be carried out in situ from a scaffold.

STEP 1 Survey the area of guttering to be repaired. Timber gutter can be purchased in various sizes and care should be taken to match the gutter as closely as possible. When viable, use existing brackets or fixings.

STEP 2 Cut out areas of rotten or damaged gutter. The cuts should be square to the face of the gutter.

STEP 3 Measure the distance of the gap in the existing gutter, then cut a piece of the new gutter, adding the length of two lapped butt joints.

STEP 4 Form a lapped butt joint on each end of the repair piece. Offer into the gap and mark the ends onto the underside of the existing gutter. Make a series of cross cuts into the underside of the existing gutter to the depth of the quirk and remove with a sharp broad chisel as you did in Step 4 in the previous task.

REED TIP

Employers will be looking at your natural talents, your workplace skills and your academic performance when deciding whether to hire you.

SASH WINDOWS

Box framed or vertical sliding sash windows were extensively used in older properties. These windows need to be maintained like any other window and may have to be repaired or replaced. Some local authorities will allow them to be replaced with modern windows that are unsympathetic to both the building and the surrounding area, while others will insist that they are replaced like for like.

Listed buildings are subject to strict controls on both the refurbishment and replacement of original features; however, a large proportion of the relevant period's housing stock is not listed but still has to be maintained. To repair or refurbish a box framed window the carpenter or joiner must have a good knowledge of how the window is constructed.

Replacing sash window cords

Often a carpenter will be asked to replace a broken sash cord. In addition to the sash cords, the sash window is made up of several components, as can be seen in Fig 8.42.

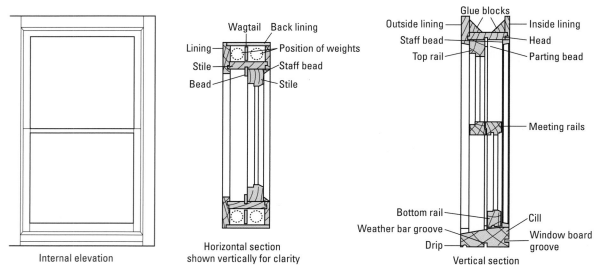

Internal elevation

Horizontal section shown vertically for clarity

Vertical section

Figure 8.42 Box-frame sliding sash window details

The actual process of replacing the sash is covered in detail in the practical that follows. Fig 8.43 shows the processes that need to be followed to re-cord the sash.

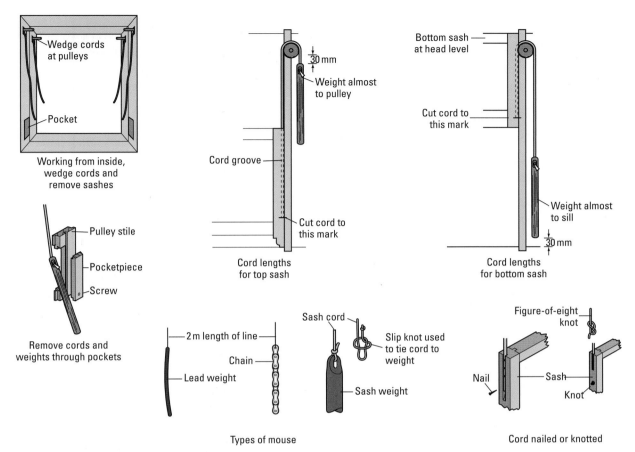

Figure 8.43 Re-cording sashes

Box sash window components

Box sash windows are actually very easy to repair. They can usually be repaired without having to take the frame out. However, within the box sash window there are a number of components that you need to be familiar with. These are outlined in Table 8.4.

Component	Description and purpose
Pocket	Pockets provide access to the sash weights and are cut into the pulley stile. These are positioned at the bottom of the box and are sometimes difficult to see after several layers of paint have been applied.
Parting bead	These separate the sashes vertically and hold them in place. Parting beads are set into a groove in the pulley stile and head of the box frame.
Staff bead	This is a moulded bead designed to hold the inner sash in place on the inside of the box frame. It is either nailed or screwed to the inside lining.
Pulley wheels	These are also referred to as axle pulleys. They are set in the pulley stile over which the sash cord runs. The wheel can be made from brass, iron or nylon and is fixed with a cast iron or brass mounting plate.
Cill	This is the bottom component member whole of the box frame. It is shaped so that rainwater is easily shed from it. It is sometimes made from hardwood, traditionally oak.
Pulley stiles	This is a side member of the box frame. It carries the weight of the sashes and their weights. It is suspended over the axle pulley or pulley wheel.

Component	Description and purpose
Balance springs	This is an alternative way of suspending sashes. There is a spiral mechanism with a spring, which is tensioned to counterbalance the weight of the sash. In more modern sash windows they are grooved into the frame stiles. In traditional windows they are set into the edges of the sash stiles.
Sash weights	These are either lead or cast iron weights. A pair counterbalances the weight of each sash.
Sliding sash	This term is used to describe the actual frame and panes of glass in a sash window.
Sash cord	These are ropes that pass over the axle pulleys (pulley wheels) and are used to hang the sashes. They are counterbalanced by sash weights.
Glazing bar	These are also known as sash bars or sash guts. They separate the panes of glass in a sash with more than one pane of glass.
Wagtails	These are softwood or plywood dividers that are suspended from the head of the boxed frame. They are located behind the stile pulley and separate the sash weights preventing them from banging into one another. They are suspended from the top within the sash weight pocket, and are not fixed rigidly so they are able to move to allow the removal of the outside weights when re-cording.

Table 8.4

DID YOU KNOW?

Most sash windows slide vertically. However there are horizontal versions called Yorkshire Lights. The sash window is believed to date back at least 250 years.

PRACTICAL TASK

7. CARRY OUT MAINTENANCE ON BOX FRAMED OR VERTICAL SLIDING SASH WINDOWS

OBJECTIVE

To be able to maintain and repair sash windows.

There are two methods of re-cording a window:

* Use a single length of sash cord to re-cord both sides of the window simultaneously.

* Re-cord each side individually using four lengths of cord.

The procedure for preparing the sashes and the frame is the same for both methods.

PPE

Ensure you select PPE appropriate to the job and site where you are working. Refer to the PPE section in Chapter 1.

TOOLS AND EQUIPMENT

Hammer

Screwdrivers

Chisels

Sash knife or pocket chisel

Utility knife

Plane

Tenon saw

Scraper

Lead mouse

Pincers

Glass paper

String

PREPARING THE FRAME

STEP 1 Start by breaking the paint joints on the inner staff bead using a utility knife.

STEP 2 Using a hammer and chisel, gently prise the vertical beads out and remove the nails with a pair of pincers by levering them out through the back of the bead so as not to damage the face. There is no need to remove the top and bottom beads.

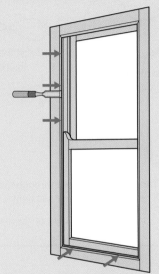

Figure 8.44 Carefully remove the nails

STEP 3 Lift the inner sash over the bottom staff bead and rest it on the window board. If the sash cord is unbroken it will need temporarily wedging to the pulley to prevent the sash weight crashing to the bottom of the sash pocket. The sash cord or cords can now be cut.

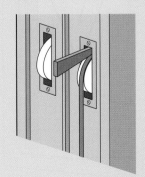

Figure 8.45 Wedge the pulley

STEP 4 The inner sash can now be lifted clear of the box frame. When working on larger sashes, this is best done with two operatives.

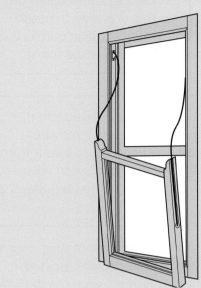

Figure 8.46 Lift inner sash clear of frame

STEP 5 Using a utility knife, break the paint joints on the parting beads and remove the beads.

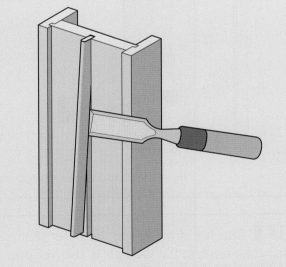

Figure 8.47 Remove the parting beads

PRACTICAL TIP

These beads are often broken during the process so it is a good idea to have some replacement beads before commencing work.

STEP 6 Nail or screw two blocks of wood about 300 mm down from the bottom of the outer sash. Allow the sash to come down and sit on the blocks. This prevents the sash dropping at one end when the sash cord is cut and prevents the sash binding in the frame.

STEP 7 If the sash cords are still attached, wedge them against the pulley. It may be necessary to break the old paint to free it from the frame. The sash cord or cords can now be cut. Lift the sash out of the frame (see Steps 2 and 3).

STEP 8 Remove the sash pockets located at the bottom of the box frame using a sash knife or chisel or a suitable wood chisel.

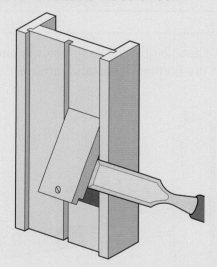

Figure 8.48 Remove the sash pockets

STEP 9 Attach a string line to the outer sash cords then remove the wedges and lower the weights to the bottom of the pocket. Do not let the weights drop in the pockets. Repeat for inner sash cords.

STEP 10 Remove any old sash cord that is still attached to both sides of the sashes. This is located in a groove down either side of the sash.

STEP 11 Any remedial work that is required on either the sash or the box frame should be carried out at this stage.

METHOD 1: RE-CORDING USING A SINGLE LENGTH OF SASH CORD

STEP 1 Attach the mouse to one end of the sash cord using a figure-of-eight loop.

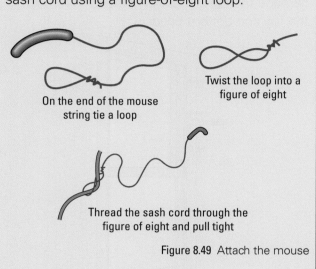

On the end of the mouse string tie a loop

Twist the loop into a figure of eight

Thread the sash cord through the figure of eight and pull tight

Figure 8.49 Attach the mouse

STEP 2 Feed the mouse over the left-hand inside pulley and out through the left-hand pocket.

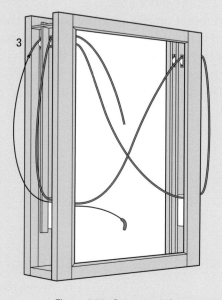

Figure 8.50 Over the left inside, out the left

STEP 3 Feed the mouse over the right-hand inside pulley and out through the right-hand pocket.

STEP 4 Feed the mouse over the left-hand outside pulley and out through the left-hand pocket.

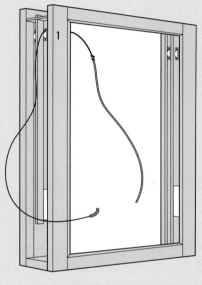

Figure 8.51 Over the left outside, out the left

STEP 5 Feed the mouse over the right-hand outside pulley and out through the right-hand pocket.

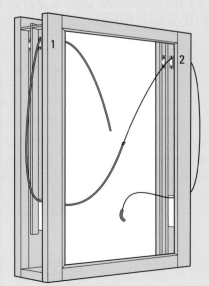

Figure 8.52 Over the right outside, out the right

STEP 6 Tie a knot in the length of sash cord hanging from the left-hand inside pulley. This will stop it from going over the pulley and back into the pocket.

STEP 7 Tie the sash weight from the right-hand pocket to the other end of the sash cord. This will prevent the sash cord being pulled back over the right-hand outside pulley.

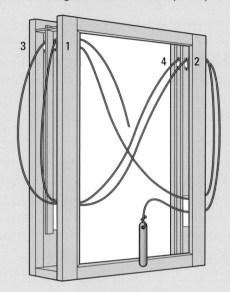

Figure 8.53 Tie on the sash weight

STEP 8 Cut the front diagonal cord running from the bottom left to the top inside right in the centre. Tie a knot in the tail hanging from the pulley and tie a sash weight to the tail hanging from the pocket.

STEP 9 Cut the middle diagonal running from bottom right to top outside left in the centre. Tie a knot in the tail hanging from the pulley and tie a sash weight to the tail hanging from the pocket.

STEP 10 Tie a knot in the tail hanging from the outside right pulley and tie a sash weight to the remaining tail.

STEP 11 Place the outside sash on the windowboard and mark the bottom of the cord groove onto the pulley stiles. Pull the sash cord until the sash weight is almost to the pulley and wedge the pulley to prevent it moving. Repeat for both sides.

STEP 12 Tie a knot in the cords in line with the marks on the frame and cut off the excess.

STEP 13 Lift the outer sash onto the window board and insert the cords into the grooves. The knot should sit in the hole provided. Nail the cords into place using round-headed nails.

STEP 14 Remove the wedges, allow the sash to sit in the frame and guide it to the head. Check that the weights are not bottoming out in the frame. Check that it clears the full length of the frame when opening and closing.

STEP 15 Replace the parting beads.

STEP 16 Measuring down from the head of the frame, mark the bottom of the inner sash cord grooves on the pulley stiles. Pull the cords until the weight clears the cill inside the pocket and wedge it. Cut the cords in line with the marks and tie a knot in the end of the cord (see Fig 8.54).

STEP 17 Remove the wedges. Pull the cords almost to the pulleys and re-wedge. Lift the inner sash onto the window board and insert the cords into the grooves. The knot should sit in the hole provided. Nail the cords into place using round-headed nails.

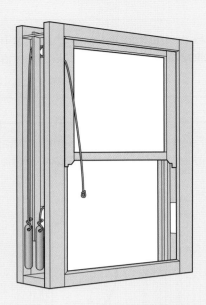

Figure 8.54 Tie a knot in the end of the cord

STEP 18 Follow Step 14 and replace or renew staff beads.

STEP 19 Replace the pockets back in the frame. Re-cord each side individually using four lengths of cord.

METHOD 2: RE-CORDING USING SEPARATE LENGTHS OF SASH CORD

STEP 1 Follow Steps 1 to 14 as above.

STEP 2 Cut four lengths of cord and feed them through the pulleys. Ensure that the cords are long enough to reach the bottom of the pocket and hang around half-way down the box frame.

STEP 3 Feed the cords over each of the pulleys using a mouse. Pull the ends out of the pocket at the bottom of the frame. Wedge the cords against the pulleys at the top.

STEP 4 Attach sash weights to the outside cords; these can be identified easily as they sit behind the wagtails. Feed the weights back into the pockets, ensuring they are sitting behind the wagtails.

STEP 5 Attach sash weights to the inside cords and feed into the pockets.

STEP 6 Repeat Steps 11 to 19 for the previous method.

REPAIRS TO BOX FRAME AND SASHES

It may be necessary to repair other components within the window. All repairs should be carried out while the sashes are out of the frame.

STEP 1 Start on the outside and work inwards. Check the condition of the cill and ensure that the mastic pointing is intact around the perimeter of the frame.

STEP 2 Check outside and inside pulleys and replace if required.

STEP 3 Having removed the parting bead, check its condition and replace if necessary.

STEP 4 Check the condition of the window board. Repair or renew as required.

STEP 5 Check both sashes, replace any broken glass, renew or repair putty pointing and repair or replace glazing bars. On occasion, it may be necessary to disassemble the sash, re-glue it and reassemble it.

STEP 6 If required, refit the sash shooting edges until optimum clearance is achieved to allow smooth operation. Take care not to remove too much material as this will allow the sash to rack in the frame, causing it to bend.

PRACTICAL TIP

Sometimes it is more cost effective to make a replacement sash. This should match the existing sash. Take care with your measurements.

STEP 7 Check the staff bead and replace it if required.

STEP 8 Treat knots, prime and stop any repairs.

PRACTICAL TIP

Knotting is the sealing of knots so they do not bleed through the finished paintwork.
Priming is the application of a suitable oil or water based base coat.
Stopping is the filling of all nail and screw holes.

REED TIP

If you want to be a good apprentice, a keen and polite attitude is essential. Listen to what people are saying, keep your eyes open, nod, smile, take it all in ... and keep your mobile phone in your pocket!

TEST YOURSELF

1. If you noticed holes in woodwork, what might this suggest?

    a. Damp

    b. Dry rot

    c. Insect infestation

    d. Wet rot

2. What term is used to describe chemical substances that can kill living organisms and spores?

    a. Biocide

    b. Virus

    c. Preservative

    d. Steriliser

3. What type of timber is more likely to be affected by the death watch beetle?

    a. New pine

    b. MDF and plywood

    c. Most softwoods

    d. Timber that is rotting or decaying

4. Which of the following types of insect treatment aims to kill off the eggs?

    a. Spraying

    b. Paste

    c. Fogging

    d. Smoke

5. What should be used before priming and painting to treat small, soft areas of wood?

    a. Wood hardener

    b. White spirit

    c. Putty

    d. All-purpose filler

6. Which of the following determines the amount of protection a preservative gives to the timber?

    a. Long-term effectiveness of the preservative

    b. Quantity that remains in the timber

    c. Depth the preservative has penetrated

    d. All of these

7. What type of brush should be used to provide a flood coat?

    a. Hard bristled

    b. Soft bristled

    c. Fine detail

    d. Emulsion brush

8. What is the name of the gutter component that directs the downpipe from the gutter, running under the soffit of the eaves and down the wall?

    a. Hopper

    b. Shoe

    c. Offset

    d. Elbow

9. What is the component that is often used to fix together sections of plastic or PVC guttering?

    a. Union piece

    b. Gasket

    c. Bolt

    d. Bracket

10. Which part of a sash window carries the weight of the sashes and their weights?

    a. Pulley wheel

    b. Staff bead

    c. Pocket

    d. Pulley stile

INDEX

A

abbreviations 50
access equipment, structural carcassing 243–4
access panels, services 184, 191
access traps, floor coverings 155–6
accident books 2, 9, 17
accident procedures 8–13
additives, foundations 72–3
adhesives 118–19
    door frames 139
agendas 65
aggregates, foundations 72
alternative building methods, sustainability 87–8
alternative energy sources, sustainability 91–3
architecture, sustainability 85–6
architraves 228–31
    scribe architraves 230–1
asbestos, Control of Asbestos at Work Regulations 3
assembly drawings 42
assembly points 34

B

barge boards, roofs 255–7, 258–61
bath panels 192
batten (exterior roof feature) 80
beam and block floors 74
biocides 282
biodegradable materials 88
biodiversity 84
biomass stoves 91
block plans 41
box sash windows 137, 303–9
building design, sustainability 85–6, 89, 93
Building Regulations 89, 93, 98
Building Regulations Part K, stairs 169
built-in frames
    door frames 138, 143–4
    window frames 143–4

C

carbon, energy sources 90
carbon footprint 85
cavity fixings 116–17
cavity walling 77
CCHP see combined cooling heat and power units
cement, foundations 72
chemical fixing 118–19
chimney openings, floors 274–5
chop saws 183
circular saws, portable power tools 114–15
cladding
    partition walls 165
    service encasements 184–92
coach bolts 117
coach screws 116
colour coded cables, electricity 31
combined cooling heat and power (CCHP) units 92
combustible materials 18, 34–6
common furniture beetle, insect infestation 284

communication 58–65
    see also information
    diversity 61–2
    documentation 62–4
    equality 61–2
    key personnel 59–60
    methods 62–5
    poor 61
    teamwork 61
    types 58–9
competent (individuals/organisations) 4
concrete
    floors 75
    formwork 75
    foundations 72
conductive materials, portable power tools 107
contamination 18, 19
Control of Asbestos at Work Regulations 3
Control of Substances Hazardous to Health Regulations (COSHH) 2, 3, 20
cornices, wall units 204
COSHH see Control of Substances Hazardous to Health Regulations
covering letters 65
creosote, maintenance 289
cutting, shaping and sanding, portable power tools 108–9
cylinder latches, door 209–10
cylinder rim locks 223–4

D

dado rails 239
damp-proof courses (DPCs) 78
damp-proof membranes (DPMs) 78
dangerous occurrences 9
death watch beetle, insect infestation 284
debris, portable power tools 107–8
dermatitis 20, 21, 32
detail drawings 44
diseases 9
documentation 62–4
    see also communication; information
door and window openings, partition walls 166
door casings 133–4
door frames 132–4
    adhesives 139
    built-in frames 138, 143–4
    fixed-in frames 137–8, 141–3
    fixing 137–9, 140
door linings 144–6
doors 204–25
    cylinder latches 209–10
    cylinder rim locks 223–4
    fire doors 133–4, 225
    fitting 212–15
    hanging 218
    hinges 206–8, 215–17
    ironmongery 206–11
    kitchen units 204
    letter plates 211, 224–5
    locks/latches 208–11, 219–24
    maintenance 290–2
    mortise locks 208–11, 219–22
    troubleshooting 218
double pitched roofs 79
DPCs see damp-proof courses
DPMs see damp-proof membranes
drawings 41–4
    equipment 50
    scales 51

drill bits, portable power tools 102, 106–7
drilling, portable power tools 115–19
dry lining 129
dry rot 282

E

ear defenders 20, 21, 32
eaves, roofs 255–8, 259–61
ECOSHH Regulations, noise 20
electric mitre saws 183
electric routers 183
electricity 28–31
    colour coded cables 31
    dangers 29–30
    Portable Appliance Testing (PAT) 28–9
    precautions 28–9
    voltages 30, 31
emergency procedures 8–13, 34–6
energy efficiency, sustainability 89–93
energy ratings 93
environment 85
    see also sustainability
estimates, job 58, 63
estimating quantities of resources 51–8
eye protection 32

F

fascia (exterior roof feature) 80
fastenings, portable power tools 115–19
fire doors 133–4, 225
fire extinguishers 35–6
fire procedures 34–6
firings (exterior roof feature) 80
first aid 12–13
first angle projection (orthographic) 44
first fixing operations 121–32
    fixings 127–30
    materials 127–30
    site datum points 122–3
    tools and equipment 123–7
fixed-in frames
    door frames 137–8, 141–3
    window frames 141–3
flat roof decking 146–56
    see also floor coverings
    installing 151
flat roofs 78, 79
floating floors 74
floor coverings 146–56
    access traps 155–6
    fixing 148–9
    installing 151
    materials 147
    noggins 149
    services openings 150
    tongue and groove chipboard floors 152–3
    tongue and groove softwood floors 154
floor joists 262–77
    ground floors 271–5
    installing 270–7
    joist positions 267–9
    openings 265–6
    strutting 266–7, 276–7
    supporting joists 263–5
    types 262–3
floor units 192–200
floors 73–5
    alternative building methods 88
    chimney openings 274–5
    concrete 75
    ground floors 73–4, 271–5

openings 265–6, 274–5
  suspended timber flooring 269–70
  upper floors 75, 275–7
formulae, estimating quantities of resources 51–6
formwork, concrete 75
foundations 68–73
  additives 72–3
  aggregates 72
  cement 72
  concrete 72
  materials 72–3
  pad foundations 70
  pile foundations 70, 72
  purpose 68
  raft foundations 70, 71
  reinforcement 73
  selecting 71–2
  shear failure 68
  stone, artificial/natural 73
  strip foundations 69, 71
  subsoils 71
  types 69–70
  water 72
frame assembly, portable power tools 112–13
framing anchors 160

G
gable end roofs 79, 246–8
gable ladders, roofs 258–9
gables, roofs 255–8
geothermal ground heat 92
goggles 32
ground floors 73–4, 271–5
grounds, timber 185–7
guttering
  components 294–5
  downpipe 301
  identifying damage 296
  joints 296
  maintenance 293–301
  materials 294
  plastic (PVC) 298–9
  profiles 295–6
  replacing 293–301
  timber 300–1

H
hand protection 32
handling materials 22–6
HASAWA see Health and Safety at Work etc. Act
haunches, portable power tools 112
hazards 5
  creating 17–18
  identifying 13–18
  maintenance 280
  method statements 14–15
  portable power tools 96–7, 107–8
  reporting 16–17
  risk assessments 14–15
  structural carcassing 242–3
  types 15–16
head protection 32
Health and Safety at Work etc. Act (HASAWA) 2, 5
Health and Safety Executive (HSE) 6, 7
health risks 21
hearing protection 20, 21, 32
heat sink systems 91

height, working at 4, 16, 17
  equipment 26–7, 242–4
  maintenance 289
  structural carcassing 242–4
hinges, door 206–8, 215–17
hipped end roofs 79
holding devices 182
house longhorn beetle, insect infestation 284
housekeeping 14
HSE see Health and Safety Executive
HVAC (Heating, Ventilation and Air-conditioning) 83, 84
hygiene 18–21

I
improvement notices 6
inaccurate estimates 58
information 40–51
  see also communication
  abbreviations 50
  conformity 49
  drawing equipment 50
  drawings 41–4
  manufacturers' technical information 47
  organisational documentation 48
  plans 41–4
  policies 46
  procedures 46
  programmes of work 44–5
  scales 51
  schedules 47
  specifications 46, 49–50
  training and development records 49
  types 40
infrastructure 83
injuries 7, 9, 10
insect infestation
  common furniture beetle 284
  death watch beetle 284
  house longhorn beetle 284
  maintenance 282–5
  powder post beetle 284
  treatment 284–5
  woodworm 283
insulation 93
internal walls 77–8
interpreting specifications 49–50
interviews 82
ironmongery, door 206–11
isometric projection 44

J
jig saws, portable power tools 112
joist coverings 146–56
  see also floor coverings

K
kitchen units 192–204
kitchen worktops 201–3

L
labour rates 57
ladders 26–7
landfill 84
latches/locks, door 208–11, 219–24
lean-to roofs 79
legislation
  health and safety 2–8
  personal protective equipment (PPE) 33
  portable power tools 103–4
leptospirosis 20, 21

letter plates, door 211, 224–5
letters, covering 65
lifting, safe 22–3
location drawings 41–2
locks/latches, door 208–11, 219–24

M
maintenance 279–310
  creosote 289
  doors 290–2
  dry rot 282
  guttering 293–301
  hazards 280
  height, working at 289
  insect infestation 282–5
  paint systems, timber 286–8
  portable power tools 102–5
  preservatives, timber 288–9
  schedules 280–1
  specifications 280–1
  timber 281–92
  timber doors 290–2
  timber, repairing 285–92
  wet rot 282
  window frames 290–2
major injuries 7, 10
Manual Handling Operations Regulations 4
manufacturers' instructions, portable power tools 102–3
manufacturers' technical information 47
mark-up 58
marking gauges 182
materials, purchasing systems 57
MDF (medium density fibreboard) 129
measurements, estimating quantities of resources 51–6
meetings 65
method statements, risk assessments 14–15
mitring, skirting 233–5
mono-pitch roofs 79
mortise holes, portable power tools 109–10
mortise locks 208–11, 219–22
mouldings, timber 226–7
MRMDF (moisture-resistant medium density fibreboard) 184

N
nailed butt joints 160
nails 117–18
near misses 5, 10, 11
node points, roofs 248
noggins, floor coverings 149
noise 20
non-conductive materials, portable power tools 107

O
orbital sanders, portable power tools 113–14
Ordnance Survey Benchmark (OSBM) 123
organic substances 88
organisational documentation 48
orthographic projection (first angle) 44
OSBM see Ordnance Survey Benchmark
over 7-day injuries 7

P
pad foundations 70
paint systems, timber 286–8
partition walls 77–8, 157–66
  cladding 165
  corner trapping plasterboard 165

door and window openings 166
erecting 161–6
fixing 160
materials 158
services 159
PAT *see* Portable Appliance Testing
performance reviews 65
perpendicular 78, 79
personal hygiene 20–1
Personal Protection at Work Regulations 4
portable power tools 103
personal protective equipment (PPE) 4, 31–3
legislation 33
portable power tools 102
picture rails 239
pile foundations 70, 72
pipe casing 188–9
pivot windows 135–6
planers, portable power tools 110
plans 41–4
plasterboard 129, 130
plastic plugs 115–16
plastic (PVC) guttering 298–9
plinths, kitchen units 203
policies 46
Portable Appliance Testing (PAT)
electricity 28–9
portable power tools 104
portable power tools 95–120
circular saws 114–15
conductive materials 107
cutting, shaping and sanding 108–9
damage 106–7
debris 107–8
drill bits 102, 106–7
drilling 115–19
fastenings 115–19
faults/defects checking 104
frame assembly 112–13
haunches 112
hazards 96–7, 107–8
jig saws 112
legislation 103–4
maintenance 102–5
manufacturers' instructions 102–3
mortise holes 109–10
non-conductive materials 107
orbital sanders 113–14
Personal Protection at Work Regulations 103
personal protective equipment (PPE) 102
planers 110
Portable Appliance Testing (PAT) 104
power sources 97
preparing 104–5
Provision and Use of Work Equipment
Regulations (PUWER) 103
radial arm saws 111
rebates 110
risk assessments 97
routers 109–10
tenons 111
transformers 98
types 98–102
waste control 107–8
working area 107–8
powder post beetle, insect infestation 284
power sources, portable power tools 97
power tools, portable *see* portable power
tools
PPE *see* personal protective equipment

preservatives, timber 288–9
procedures 46
profiles 132
guttering 295–6
timber frames 139
programmes of work 44–5
prohibition notices 6
proofing, timber frames 139
protective clothing 32
Provision and Use of Work Equipment
Regulations (PUWER) 3–4
portable power tools 103
purchasing systems, materials 57
purlins (exterior roof feature) 80
PUWER *see* Provision and Use of Work
Equipment Regulations

Q
quantities of resources, estimating 51–8
quotes 58, 63

R
radial arm saws, portable power tools 111
raft foundations 70, 71
rawl bolts 116
rebates
door frames 134
portable power tools 110
regulations, health and safety 2–8
reinforcement, foundations 73
repairing timber 285–92
Reporting of Injuries, Diseases and
Dangerous Occurrences Regulations
(RIDDOR) 2, 8, 9–10
resources
estimating quantities 51–8
finite/renewable 85–8
respiratory protection 32, 33
reveals 132
RIDDOR *see* Reporting of Injuries, Diseases
and Dangerous Occurrences Regulations
ridge (exterior roof feature) 80
risk assessments 14–15
portable power tools 97
risks 5
health risks 21
roofs 78–82
alternative building methods 88
barge boards 255–7, 258–61
coverings 80–2
eaves 255–8, 259–61
exterior features 80
gable end roofs 79, 246–8
gable ladders 258–9
gables 255–8
node points 248
timber bracing 248
traditional roofing 246–8
truss rafter roofs 244–6, 249–54
types 78–9, 88
verges 255–8, 259–61
wall plates 247–8
routers, portable power tools 109–10

S
safety notices 36–7, 65
sash windows 302–9
box sash windows 137, 303–9
cords 302–3
sliding sash windows 134–7
scaffold 26–7

scales, drawings 51
schedules 47
screws 116
scribe architraves 230–1
scribing internal corners, skirting 236–7
SDS (special direct system) drills 99
second fixing operations, tools and
equipment 182–3
sectional drawings 42–3
services 83
access panels 184, 191
detecting/protecting 193
locating 119
partition walls 159
service encasements and cladding 184–92
service providers 83
services openings, floor coverings 150
shear failure, foundations 68
signs 36–7, 65
site datum points 122–3
site plans 41
skew nailing 160
skirting 231–8
mitring 233–5
scribing internal corners 236–7
sliding sash windows 134–7
solar thermal hot water system 91
solid floors 74
solid masonry walls 77
specifications 46
information 46, 49–50
interpreting 49–50
maintenance 280–1
spreader bars 252
stairs 167–79
assembling 169–73
Building Regulations Part K 169
erecting 174–5
fixing 178–9
installing 175–8
levelling 174–5
terms 167–9
stone, artificial/natural, foundations 73
storing materials 22–6, 31
storm proof windows 134–5
strip foundations 69, 71
structural carcassing 241–4
see also floors; roofs
access equipment 243–4
hazards 242–3
height, working at 242–4
strutting, floor joists 266–7, 276–7
sub-contractors 6
subsoils, foundations 71
suspended timber flooring 269–70
sustainability 84–93
alternative building methods 87–8
alternative energy sources 91–3
architecture 85–6
building design 85–6, 89, 93
Building Regulations 89
energy efficiency 89–93
insulation 93
resources, finite/renewable 85–8

T
tank stands 254
technical information, manufacturers' 47
tenders 58, 63
tenons, portable power tools 111

terms
  stairs 167–9
  timber frames 139
tiles (exterior roof feature) 80
timber, maintenance 281–92
timber bracing, roofs 248
timber defects 130–2
timber doors, maintenance 290–2
timber floors 74
timber framed walls 77
timber frames 132–46
  see also door frames; window frames
  construction 139
  installing 139–40
  internal/external position 139
  profiles 139
  proofing 139
  size 139
  terms 139
  wood 139
timber guttering 300–1
timber, repairing 285–92
timber stud partition walls *see* partition walls
tongue and groove chipboard floors 152–3
tongue and groove softwood floors 154

toolbox talks 7, 8, 19
tools, portable power *see* portable power tools
traditional roofing 246–8
traditional windows 134–5
training and development records 49
transformers, portable power tools 98
trestle props 252
truss rafter roofs 244–6, 249–54

U
upper floors 75, 275–7
utilities 83
  see also services

V
VAT (Value Added Tax) 63, 64
verges, roofs 255–8, 259–61

W
wall plates, roofs 247–8
wall units 192–200, 203–4
walls 75–8
  cavity masonry 77
  internal 77–8
  partition 77–8, 157–66
  solid masonry 77

timber framed 77
waste control 25–6
  portable power tools 107–8
water, foundations 72
water tank stands 254
welfare facilities 18–19
wet rot 282
wind turbines 92
window frames
  built-in frames 143–4
  fixed-in frames 141–3
  maintenance 290–2
  types 134–7
windows
  pivot windows 135–6
  sash windows 134–7, 302–9
  storm proof windows 134–5
  traditional windows 134–5
woodworm, insect infestation 283
Work at Height Regulations 4
  see also height, working at
work programmes 44–5
working area, portable power tools 107–8
working platforms 26–7
worktops 195–7, 201–3

ACKNOWLEDGEMENTS

The publisher would like to thank David Wilkins for his review and feedback on this book.

The author and the publisher would also like to thank the following for permission to reproduce material:

Images and diagrams

Alamy: Arcaid Images: 3.15, blickwinkel: 3.16, Mike Booth: 8.2, Nigel Cattlin: 8.7, Peter Davey: chapter 1 opener, russ witherington: 5.33, Scott Camazine: 8.8; **BSA**: 2.4; **Energy Saving Trust © 2013**: 3.31; **Fotolia**: 1.1, 1.2, 1.3, 1.5, 1.6, 1.7, 1.8, 1.14, 1.15, 1.16, 2.26, 8.3, 8.5; **Helfen**: 2.3; **instant art**: table 1.15; **Image courtesy of Robert Bosch Ltd**: 6.17; **iStockphoto**: 1.11, 2.27, 3.7, 3.19, 3.21, 3.22, 3.24, 3.25, 3.28, 3.29, 3.30, 4.1, 5.7, 5.32, 5.36, 5.41, 8.27, 8.28; **Leeds College of Building**: 5.73, 5.74, 5.75, 5.76; **Nelson Thornes**: 1.9, 1.10, 1.12, 1.13, 4.4, 4.8, 4.9, 4.10, 4.11, 4.12, 4.13, 4.14, 4.15, 4.16, 4.17, 4.18, 4.19, 4.20, 4.21, 4.22, 4.23, 4.24, 4.25, 4.26, 4.27, 4.28, 4.30, 4.31, 4.32, 4.34, 4.35, chapter 5 opener, 5.3, 5.4, 5.5, 5.6, 5.8, 5.9, 5.10, 5.11, 5.13, 5.14, 5.15, 5.17, 5.18, 5.19, 5.20, 5.22, 5.23, 5.25, 5.26, 5.28, 5.29, 5.30, 5.31, 5.57, 5.58, 5.59, 5.60, 5.61, 5.97, 5.98, 5.99, 5.100, 5.101, 5.104, 5.105, 5.106, 5.107, chapter 6 opener, 6.1, 6.2, 6.29, 6.30, 6.31, 6.32, 6.33, 6.34, 6.35, 6.36, 6.52, 6.53, 6.54, 6.55, 6.56, 6.57, 6.58, 6.59, 6.60, 6.61, 6.62, 6.63, 6.64, 6.65, 6.66, 6.67, 6.68, 6.69, 6.70, 6.71, 6.72, 6.73, 6.74, 6.75, 6.76, 6.77, 6.78, 6.79, 6.80, 6.81, 6.82, 6.83, 6.84, 6.85, 6.86, 6.87, 6.88, chapter 7 opener, chapter 8 opener, 8.16, 8.17, 8.18, 8.19, 8.20, 8.21, 8.22, 8.23, 8.24; **Peter Brett**: 2.2, 2.5, 2.6; **Reprinted with permission from WOODWEB - The Information Resource for the Woodworking Industry - www.woodweb.com**: 5.39; **Science Photo Library**: Peter Gardiner: 1.4; **Shutterstock**: chapter 2 and 3 opener, 3.14, 3.17, 3.23, 3.26, 3.27, chapter 4 opener, 4.2, 4.3, 4.29, 4.33, 5.2, 5.21, 5.24, 5.27, 5.34, 5.35, 5.37, 5.38, 5.40, 6.51, 8.1, 8.9, 8.10, 8.11, 8.12, 8.13, 8.15; © **Victoria and Albert Museum, London**: 5.93; **Wikipedia**: 3.18, 8.6.

Every effort has been made to trace the copyright holders but if any have been inadvertently overlooked the publisher will be pleased to make the necessary arrangements at the first opportunity.